计算机类技能型理实一体化新形态系列

Spring框架应用开发
——基于Spring Boot

（微课视频版）

主编 张寺宁 吴边

U0223267

清华大学出版社
北京

内 容 简 介

本书以 Spring Boot 为基础介绍 Spring 框架的应用开发，适合 Spring 应用开发的初学者使用。全书首先介绍了 Spring 相关知识点，进而引出 Spring Boot 的实践应用，包括 Spring Boot 数据操作、定时任务、消息队列、前后端开发、安全控制和项目部署等内容。全书采用项目式教学模式，以项目、任务为驱动讲解 Spring Boot 的理论知识和实践应用。本书为立体化、活页式教材，配套 PPT、源代码、视频资源二维码、活页式综合案例、课后习题解答等电子资源，使读者能够更加灵活、方便地进行学习。

本书既可作为高等院校计算机相关专业的教材，也可作为 Spring 应用开发编程爱好者的自学书籍。

本书封面贴有清华大学出版社防伪标签，无标签者不得销售。
版权所有，侵权必究。举报：010-62782989，beiqinquan@tup.tsinghua.edu.cn。

图书在版编目(CIP)数据

Spring 框架应用开发：基于 Spring Boot：微课视频版/张寺宁，吴边主编. —北京：清华大学出版社，2024.6

（计算机类技能型理实一体化新形态系列）
ISBN 978-7-302-66104-7

Ⅰ．①S… Ⅱ．①张… ②吴… Ⅲ．①JAVA 语言—程序设计 Ⅳ．①TP312.8

中国国家版本馆 CIP 数据核字(2024)第 081996 号

责任编辑：张龙卿
封面设计：刘代书　陈昊靓
责任校对：李　梅
责任印制：沈　露

出版发行：清华大学出版社
网　　址：https://www.tup.com.cn，https://www.wqxuetang.com
地　　址：北京清华大学学研大厦 A 座　　邮　编：100084
社 总 机：010-83470000　　邮　购：010-62786544
投稿与读者服务：010-62776969，c-service@tup.tsinghua.edu.cn
质量反馈：010-62772015，zhiliang@tup.tsinghua.edu.cn
课件下载：https://www.tup.com.cn，010-83470410

印 装 者：三河市君旺印务有限公司
经　　销：全国新华书店
开　　本：185mm×260mm　　印　张：19　　字　数：458 千字
版　　次：2024 年 7 月第 1 版　　印　次：2024 年 7 月第 1 次印刷
定　　价：59.00 元

产品编号：102064-01

前　言

为了帮助读者学习、掌握和使用 Java 语言设计并了解开发项目的方法，编者携手企业有经验的工程师开发了一整套 Java 技术体系丛书。本丛书共 5 本，包括《Java 面向对象程序设计（微课视频版）》《Java Web 程序设计（微课视频版）》《Spring 框架应用开发——基于 Spring Boot（微课视频版）》《Spring Cloud 微服务应用开发——基于 Alibaba Nacos（微课视频版）》《Spring 微服务系统部署（微课视频版）》。

本书介绍了 Spring 框架以及 Spring Boot 核心技术。全书为立体化、活页式教材，采用项目式教学模式，以项目、任务为驱动讲解 Spring Boot 理论知识和实践应用，并配套相应的电子资源。

全书共 11 个项目，项目 1 介绍了 Spring 的核心概念和编程实现。项目 2 介绍了 Bean 的核心概念和编程实现；项目 3 介绍了 Spring AOP 的核心概念和编程实现；项目 4 介绍了 Spring Boot 的核心概念和编程实现；项目 5 介绍了如何在 Spring Boot 项目中使用 Mybatis 框架和数据库连接池读写数据库数据并进行事务控制；项目 6 介绍了使用 Cron 表达式进行 Spring Task 定时任务编程；项目 7 介绍了消息队列中间件的概念和 Kafka 消息队列编程；项目 8 介绍了 Spring Boot 集成 Spring MVC 进行 Web 应用的后端编程；项目 9 介绍了 Spring Boot 前后端不分离项目和前后端分离项目的前端编程实现；项目 10 介绍了 Spring Boot 项目如何使用 Security 组件进行资源访问的认证和授权；项目 11 介绍了 Spring Boot 前后端不分离项目和前后端分离项目的部署过程。

本书建议授课学时为 84 个学时。

本书项目 1 和项目 2 由吴边编写，项目 3～10 由张寺宁编写，项目 11 由古凌岚编写。本书配有详细的 PPT 讲义、教学视频、本书源代码、课后练习解答等电子资源，这些电子资源可从清华大学出版社官网下载。

由于编者水平有限，不足之处在所难免，敬请广大读者批评、指正。

编　者
2024 年 1 月

目 录

项目1 初识 Spring ·· 1

任务1.1 了解 Spring ·· 1
1.1.1 Spring 简介 ·· 1
1.1.2 Spring 的作用 ·· 1
1.1.3 Spring 的发展 ·· 2

任务1.2 认识 Spring 项目模板 ·· 3

任务1.3 了解 Spring 容器 ·· 4
1.3.1 BeanFactory ·· 4
1.3.2 ApplicationContext ·· 4

任务1.4 体验 Spring 编程 ·· 5
1.4.1 环境准备 ·· 5
1.4.2 创建 Spring Maven 项目 ·· 5
1.4.3 认识注解 ·· 6
1.4.4 基于注解方式的 Spring 编程 ·· 7

任务1.5 了解 Spring 控制反转(IOC) ·· 10
1.5.1 初识 IOC ·· 11
1.5.2 依赖注入 ·· 14

任务1.6 综合案例:获取 Spring 中 Bean 的相关信息 ·· 16
1.6.1 案例任务 ·· 16
1.6.2 任务分析 ·· 16
1.6.3 任务实施 ·· 16

小结 ·· 20

课后练习:获取 Spring Bean 对象相关信息并过滤 ·· 20

项目2 认识 Spring 中的 Bean ·· 21

任务2.1 基于注解创建无变量属性 Bean ·· 21
2.1.1 通过@Bean 标识方法创建 Bean ·· 21
2.1.2 通过@ComponentScan 自动扫描方式创建 Bean ·· 25
2.1.3 通过@Import 创建 Bean ·· 26
2.1.4 通过 FactoryBean 工厂创建 Bean ·· 30

任务2.2 基于注解创建有变量属性的 Bean ·· 32

2.2.1 利用有参构造方法注入	32
2.2.2 利用 set 方法注入	33
2.2.3 利用注解注入	35

任务 2.3 了解 Bean 的作用域 41
    2.3.1 初识 Bean 作用域 41
    2.3.2 Bean 的作用域与线程安全 42
    2.3.3 Spring 中单例 Bean 的多线程访问控制 43

任务 2.4 了解 Bean 的生命周期 43

任务 2.5 综合案例：统计用户登录次数 44
    2.5.1 案例任务 44
    2.5.2 任务分析 44
    2.5.3 任务实施 44

小结 48

课后练习：校验并分类统计登录用户信息 48

项目 3 Spring AOP 编程 49

任务 3.1 了解代理机制 49

任务 3.2 初识 Spring AOP 50
    3.2.1 AOP 简介 50
    3.2.2 AOP 术语 51
    3.2.3 Spring 的两种 AOP 实现 52

任务 3.3 基于注解的 AOP 编程 57

任务 3.4 综合案例：利用 AOP 实现访问控制 62
    3.4.1 案例任务 63
    3.4.2 任务分析 63
    3.4.3 任务实施 63

小结 66

课后练习：利用 AOP 方法实现权限认证 66

项目 4 初识 Spring Boot 67

任务 4.1 了解 Spring Boot 67

任务 4.2 体验 Spring Boot 编程 67
    4.2.1 创建 Spring Boot 项目 68
    4.2.2 分析项目结构及 pom.xml 文件 68
    4.2.3 运行 Spring Boot 项目并打包 71
    4.2.4 设置 Spring Boot 服务开机启动 73

任务 4.3 体验 Spring Boot 单元测试 75
    4.3.1 使用默认测试类进行单元测试 75
    4.3.2 手动创建测试类进行单元测试 76

- 任务 4.4　了解 Spring Boot 配置文件 ····· 78
  - 4.4.1　初识 yaml 语法 ····· 78
  - 4.4.2　读取 yaml 中的配置 ····· 79
- 任务 4.5　Spring Boot 多环境配置 ····· 85
  - 4.5.1　基于多文件的多环境配置 ····· 85
  - 4.5.2　基于单文件的多环境配置 ····· 86
- 任务 4.6　综合案例：用 Spring Boot 实现基于 TCP 服务的请求响应 ····· 87
  - 4.6.1　案例任务 ····· 87
  - 4.6.2　任务分析 ····· 87
  - 4.6.3　任务实施 ····· 88
- 小结 ····· 91
- 课后练习：用 Spring Boot 实现基于 TCP 服务网购功能 ····· 91

项目 5　Spring Boot 数据操作和事务处理 ····· 92

- 任务 5.1　初识数据库连接池 ····· 92
  - 5.1.1　Hikari 连接池 ····· 92
  - 5.1.2　Druid 连接池 ····· 94
- 任务 5.2　Spring Boot Mybatis 数据操作 ····· 96
  - 5.2.1　Mybatis 简介 ····· 96
  - 5.2.2　Spring Boot 引入 Mybatis ····· 97
  - 5.2.3　Spring Boot 引入 Lombok 插件 ····· 98
  - 5.2.4　Mybatis 注解进行单表数据操作 ····· 99
  - 5.2.5　Mybatis 注解进行多表关联查询 ····· 104
  - 5.2.6　Mybatis 注解动态 SQL ····· 111
  - 5.2.7　Mybatis 数据缓存机制 ····· 121
- 任务 5.3　Spring Boot 事务 ····· 125
  - 5.3.1　事务简介 ····· 126
  - 5.3.2　Spring Boot 声明式事务控制 ····· 126
  - 5.3.3　事务隔离级别 ····· 127
  - 5.3.4　事务传播机制 ····· 128
  - 5.3.5　编程实现基于注解的事务控制 ····· 129
- 任务 5.4　综合案例：用 Spring Boot 模拟实现人员账户管理 ····· 130
  - 5.4.1　案例任务 ····· 130
  - 5.4.2　案例分析 ····· 131
  - 5.4.3　任务实施 ····· 131
- 小结 ····· 131
- 课后练习：用 Spring Boot 模拟实现人员账户转账 ····· 131

项目 6 Spring Boot 定时任务 ································· 132

任务 6.1 Cron 表达式和定时任务框架 ································· 132
6.1.1 初识 Cron 表达式 ································· 132
6.1.2 常用的定时任务框架 ································· 134

任务 6.2 基于 Spring Task 定时任务编程 ································· 135
6.2.1 初识 Spring Task ································· 135
6.2.2 Spring Task 基于单个定时任务编程实现 ································· 135
6.2.3 Spring Task 基于多个定时任务编程实现 ································· 139
6.2.4 Spring Task 动态定时任务编程实现 ································· 142

任务 6.3 综合案例：利用 Spring Task 实现定时闹钟 ································· 146
6.3.1 案例任务 ································· 146
6.3.2 案例分析 ································· 146
6.3.3 任务实施 ································· 147

小结 ································· 147
课后练习：定时清除过期闹钟任务 ································· 147

项目 7 Spring Boot 消息队列 ································· 148

任务 7.1 初识消息队列 ································· 148
7.1.1 消息队列简介 ································· 148
7.1.2 常用的消息队列中间件 ································· 149

任务 7.2 基于 Kafka 的消息队列编程 ································· 149
7.2.1 Kafka 简介 ································· 149
7.2.2 Kafka 安装和配置 ································· 151
7.2.3 Spring Boot 引入 Kafka ································· 151
7.2.4 Spring Boot 基于 Kafka 的编程实现 ································· 152

任务 7.3 综合案例：Kafka 采集主机运行信息 ································· 159
7.3.1 案例任务 ································· 159
7.3.2 案例分析 ································· 159
7.3.3 任务实施 ································· 160

小结 ································· 166
课后练习：Kafka 采集键盘输入字符数据 ································· 167

项目 8 Spring Boot Web 应用开发——后端 ································· 168

任务 8.1 初识 Spring MVC ································· 168
8.1.1 Spring MVC 简介 ································· 168
8.1.2 Spring MVC 工作流程 ································· 169
8.1.3 Spring Boot 引入 Spring MVC ································· 170
8.1.4 Spring MVC 单元测试工具——MockMvc ································· 170

| 任务 8.2 | Spring MVC 访问静态资源 | 171 |

任务 8.3 Spring MVC 访问动态资源——映射请求 ································· 172
    8.3.1 @Controller 注解 ·· 172
    8.3.2 @RequestMapping 注解 ·· 172
    8.3.3 组合注解 ·· 175

任务 8.4 Spring MVC 访问动态资源——获取请求数据 ···························· 175
    8.4.1 @RequestParam 注解 ·· 175
    8.4.2 @RequsetBody 注解 ·· 180
    8.4.3 @PathVariable 注解 ·· 184

任务 8.5 Spring MVC 访问动态资源——输出响应 ································· 186
    8.5.1 跳转页面 ·· 186
    8.5.2 回写数据 ·· 191

任务 8.6 Spring MVC Restful 风格编程 ··· 192
    8.6.1 初识 Restful 风格 ··· 192
    8.6.2 Spring MVC 实现 Restful 风格编程 ·· 193

任务 8.7 Spring MVC 拦截器 ··· 196
    8.7.1 定义拦截器 ·· 196
    8.7.2 使用拦截器 ·· 197
    8.7.3 拦截器和过滤器 ··· 199

任务 8.8 Spring MVC 文件上传和下载 ·· 200
    8.8.1 Spring MVC 文件上传 ·· 200
    8.8.2 Spring MVC 文件下载 ·· 203

任务 8.9 综合案例：员工信息管理 ··· 204
    8.9.1 案例任务 ·· 204
    8.9.2 案例分析 ·· 204
    8.9.3 任务实施 ·· 204

小结 ··· 205
课后练习：学生信息管理 ··· 205

项目 9 Spring Boot Web 应用开发——前端 ······································· 206

任务 9.1 了解 Spring Boot Web 应用前端实现方式 ································· 206
任务 9.2 利用 JSP 模板引擎实现前端功能 ··· 207
    9.2.1 初识 JSP 模板引擎 ·· 207
    9.2.2 Spring Boot 引入并配置 JSP 模板引擎 ·································· 207
    9.2.3 编写控制器类和 JSP 前端页面实现增、删、改、查 ················ 209

任务 9.3 利用 Thymeleaf 模板引擎实现前端功能 ··································· 216
    9.3.1 初识 Thymeleaf 模板引擎 ·· 216
    9.3.2 Spring Boot 引入 Thymeleaf 模板引擎 ································· 216
    9.3.3 Thymeleaf 语法 ··· 216

| 9.3.4 编写 Thymeleaf 前端页面实现增、删、改、查 …………………… 220
| 任务 9.4 利用 Vue 实现前端功能 …………………… 223
| 9.4.1 初识 Vue …………………… 224
| 9.4.2 搭建 Vue3 开发环境 …………………… 225
| 9.4.3 创建 Vue3 项目 …………………… 225
| 9.4.4 Vue3 项目目录结构及访问机制 …………………… 228
| 9.4.5 Vue3 组件入口函数——setup 函数 …………………… 230
| 9.4.6 创建和渲染响应式数据 …………………… 233
| 9.4.7 修改响应式数据 …………………… 243
| 9.4.8 异步加载响应式数据——Axios 组件 …………………… 245
| 9.4.9 Vue3 页面跳转——Vue-Router 组件 …………………… 251
| 9.4.10 Vue3 集成 Element-Plus …………………… 259
| 任务 9.5 综合案例：基于 Vue3 实现员工信息管理 …………………… 264
| 9.5.1 案例任务 …………………… 264
| 9.5.2 案例分析 …………………… 264
| 9.5.3 任务实施 …………………… 264
| 小结 …………………… 264
| 课后练习：学生信息管理 …………………… 265

项目 10 Spring Boot 安全控制——Security …………………… 266

| 任务 10.1 初识 Spring Security …………………… 266
| 10.1.1 Security 简介 …………………… 266
| 10.1.2 Spring Boot 中引入 Spring Security …………………… 267
| 任务 10.2 Spring Security 单用户认证和授权 …………………… 267
| 10.2.1 Spring Security 默认登录注销认证 …………………… 267
| 10.2.2 Spring Security 自定义登录注销认证 …………………… 270
| 10.2.3 Spring Security 自定义授权 …………………… 273
| 10.2.4 Spring Security 静态资源的访问控制 …………………… 278
| 任务 10.3 Spring Security 多用户认证和授权 …………………… 278
| 任务 10.4 综合案例：利用 Spring Security 进行安全控制 …………………… 281
| 10.4.1 案例任务 …………………… 281
| 10.4.2 案例分析 …………………… 281
| 10.4.3 案例实施 …………………… 281
| 小结 …………………… 282
| 课后练习：前后端分离项目的安全控制 …………………… 282

项目 11 Spring Boot Web 项目部署 …………………… 283

| 任务 11.1 部署前后端不分离项目 …………………… 283
| 11.1.1 基于 Jar 项目部署 …………………… 283

11.1.2　基于 War 项目部署 ……………………………………………… 285
任务 11.2　前后端分离项目部署 ………………………………………………… 288
小结 …………………………………………………………………………………… 291
课后练习：学生信息管理项目部署 ………………………………………………… 291

参考文献 …………………………………………………………………………… 292

项目 1　初识 Spring

现代的 Java 应用开发大多是基于 Spring 的应用开发。使用 Spring 能够让开发人员可以专注于业务逻辑的开发,把代码中的基础设施交给 Spring 负责,以减少开发时间,提高编码效率。Spring 框架在 Java 应用开发中占有重要地位,是大多数 Java 开发人员必须要学习的一门技术。

任务 1.1　了解 Spring

Spring 的中文意思是春天,意味着 Spring 的出现将为 Java 开发人员带来一个舒适、便捷的开发环境。本节将对 Spring 的相关知识做初步的介绍,包括 Spring 是什么、Spring 能做什么、Spring 的发展等内容。

1.1.1　Spring 简介

Spring 是一个基于 Java EE 应用程序开发的轻量级开源框架。该框架由罗德·约翰逊(Rod Johnson)于 2002 年创建,用于解决 Java EE 编程开发中的复杂性,实现敏捷开发。Spring 使开发人员能够使用普通的 JavaBean 实现业务流程,同时为所有的 JavaBean 提供了一个统一的应用配置框架。Spring 的两大核心技术就是控制反转(inversion of control,IOC)和面向切片编程(aspect oriented programming,AOP)。

1.1.2　Spring 的作用

Spring 将从图 1-1 所示的 7 个方面助力 Java 应用开发。

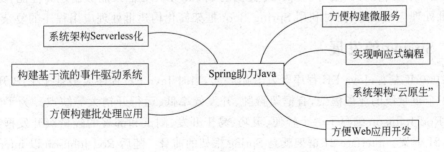

图 1-1　Spring 助力 Java

1. 方便构建微服务

微服务架构是当前 Java 开发的主流。微服务使得应用程序代码能够以可管理的、独立的模块化形式交付，同时各模块业务功能相对隔离，能够使应用程序更容易维护。Spring 提供的 Spring Boot 框架使开发人员能够更方便、更快捷地构建微服务应用程序，另外，提供的 Spring Cloud 框架可以方便管理和监控各个微服务。

2. 实现响应式编程

响应式编程是一种基于数据流的异步非阻塞编程模型，其本质是对数据流或某种变化所作出的反应。在使用时，开发人员能利用响应式编程的背压机制更方便地进行流量控制。在高并发场景下，能够更好地利用多核心 CPU 的能力构建弹性的消息驱动系统。

3. 系统架构"云原生"

"云原生"的架构不仅能够使开发完毕的系统更方便地迁移上云，而且使系统中的非业务代码和业务代码最大化松耦合，对于系统中的非业务功能则交给 Spring 管理。Spring 提供的 Spring Cloud 框架负责应用程序在云中运行的许多非功能业务，如网关 API、服务配置、服务注册发现、负载均衡、状态跟踪、熔断机制等。

4. 方便 Web 应用开发

Spring 提供的许多子项目可以方便构建 Web 应用程序。例如，Spring 提供的 Spring Boot 框架可以简化 Web 应用程序的配置。提供的 Spring Security 框架集成了许多行业标准的身份验证协议，增强了 Web 应用程序的安全性。提供的 Spring Data 框架便于 Web 应用程序访问关系数据库和非关系数据库。

5. 系统架构 Serverless 化

Serverless 即无服务，无服务不是没有服务器，只是将服务器的运维、管理和分配都托管给了云提供商，开发人员只需关注业务逻辑代码的开发，而不必关注底层计算、存储资源的使用和运维。Spring 产品组合为构建 Serverless 系统提供了强大的功能集合，便于开发人员构建 Serverless 化的系统架构。

6. 构建基于流的事件驱动系统

事件驱动系统采用基于事件的异步数据收发模式，这样能够减少数据收发两端不必要的同步等待时间，提升系统效率。Spring 提供的 Spring Cloud Stream 框架整合了诸如 Kafka、RabbitMQ 等主流的流式消息组件，简化了构建事件驱动系统的复杂性。

7. 方便构建批处理应用

批处理模式就是系统无须与用户交互即可自动、高效地处理大量信息，如定时汇总计算、定时通知、定时任务调度等。Spring 提供的 Spring Batch 框架能够创建高性能、可扩展、弹性的批处理应用程序，同时利用 Spring Boot 框架简化构建批处理应用程序的复杂度。

1.1.3 Spring 的发展

在 2000 年左右，Java EE 程序开发开始兴起，当时 Java EE 程序大多基于 EJB 开发，而 EJB 是一个重量级的业务框架，有框架臃肿、开发效率低、运行开销大等缺点。为了解决上述问题，Rod Johnson 编写了一个轻便、灵巧、易于开发、测试和部署的轻量级开发框架——interface 21 框架。interface 21 框架就是 Spring 框架的前身。随后 Rod Johnson 以 interface 21 框架为基础，经过重新设计于 2004 年 3 月 24 日正式发布了 Spring 1.0 版，Spring 正式进入公众视野。Spring 的整个发展流程如图 1-2 所示。

目前，Spring 已经发展到了 Spring 6 版本。Spring 5 到 Spring 6 是一次大版本号升级，

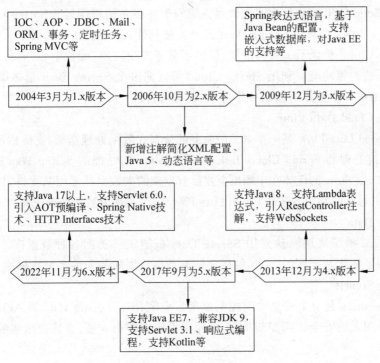

图 1-2　Spring 发展历程

这次升级是阻断式的,加入了很多新特性,同时不向下兼容。Spring 官方将 Spring 6 称为 Spring 下一个十年的新开端。

任务 1.2　认识 Spring 项目模板

Spring 构建了一系列项目模板,便于我们搭建自己的应用程序。下面介绍一些常用的项目模板。

1. Spring Boot

Spring 是一个大而全的框架。随着 Spring 整合了越来越多的框架和组件,Spring 本身变得越来越复杂,配置过多,容易让初学者陷入"配置灾难"。因此,Spring 推出了 Spring Boot 框架,Spring Boot 采用约定优于配置的原则,简化了 Spring 配置和应用程序部署,降低了 Spring 框架使用门槛。Spring Boot 可以基于最少配置创建独立的 Spring 应用程序,实现应用一键启动和部署。一个 Spring Boot 应用就可以看作一个单体的微服务项目。

2. Spring Cloud

当前主流软件架构大多是基于云原生的分布式架构,一个系统会根据业务功能做模块拆分。每个业务功能模块作为一个单独的服务,可以使用 Spring Boot 框架来开发,不同的服务之间进行相互调用。部署时,不同的服务将分布式部署在不同的服务器上。但是随着服务数量的越来越多,服务之间的调用和管理也越来越复杂,需要定制一套行之有效的框架来开发和管理分布式服务。

Spring Cloud 的出现就是为解决微服务架构中的服务治理问题。Spring Cloud 完全基于 Spring Boot 开发,它利用 Spring Boot 的开发便利性,简化了分布式服务的开发、配置和部署。Spring Cloud 整合了一系列框架,提供了服务发现注册、配置中心、消息总线、负载均衡、熔断、数据监控等功能。利用 Spring Cloud 可以把多个 Spring Boot 单体微服务项目统一管理起来,组成一个整体的微服务集群,实现有效管理。

3. Spring Cloud Data Flow

Spring Cloud Data Flow 是一个基于微服务的分布式数据处理框架,支持构建基于 Spring Cloud Stream 流处理和 Spring Cloud Task/Spring Batch 批处理的 Spring Boot 微服务项目。Spring Cloud Data Flow 的优势在于如果企业没有专业的大数据技术团队支持,只要熟悉 Java 和 Spring Boot,也能够利用 Spring Cloud Data Flow 编程实现大数据处理任务。

4. Spring Data

Spring Boot 框架底层默认采用 Spring Data 框架统一访问各种数据库。Spring Data 的作用就是为了简化数据库的访问,包括常见关系型数据库和非关系型数据库。

5. Spring Security

Spring Security 是一个安全访问控制框架。它利用了 Spring IOC 和 AOP 思想,为应用系统提供声明式的安全访问控制功能,使应用系统能够更方便、更快捷地实现安全控制。

任务 1.3　了解 Spring 容器

Spring 容器是 Spring 框架的核心。Spring 利用容器来帮助我们管理普通的 JavaBean 对象(在后续内容中 JavaBean 对象统一简称为 Bean)。Spring 提供了两种类型的容器:一种是 BeanFactory;另一种是 ApplicationContext。下面分别对两者做介绍。

1.3.1　BeanFactory

BeanFactory 是访问 Spring 容器的根接口,该接口为 Spring 依赖注入功能提供支持。BeanFactory 接口在 org.springframework.beans.factory.BeanFactory 包中定义。BeanFactory 提供了基本的容器功能,包括实例化 Bean 和获取 Bean。BeanFactory 在应用启动时不会去实例化 Bean,只有程序访问 Bean 的时候才会去实例化。这样能够使应用启动时占用较少资源,但是获取 Bean 的速度就会相对慢一些,因为临时创建 Bean 也需要时间。Spring 中有很多实现类和子接口继承了 BeanFactory 接口,因此开发中很少直接使用,而多使用其子类。

1.3.2　ApplicationContext

ApplicationContext 是 BeanFactory 的子接口,是开发中常用的 Spring 容器。它在 org.springframework.context.ApplicationContext 包中定义。与 BeanFactory 不同的是,ApplicationContext 在启动时就会实例化所有的 Bean,同时还可以选择性地为某些 Bean 配置懒加载来延迟 Bean 的实例化。此外,ApplicationContext 在 BeanFactory 的所有功能基础上还添加了对国际化、资源访问、事件传播、Web 应用等方面的支持。

实例化 ApplicationContext 容器方法如下：

```
ApplicationContext ctx = new AnnotationConfigApplicationContext(SpringDemoConfig.class);
```

其中，AnnotationConfigApplicationContext 类会从类路径 ClassPath 中寻找标识为 @Configuration 的 Spring 配置类，用于实例化 ApplicationContext 容器。

任务 1.4 体验 Spring 编程

目前，Java 开发的主流工具是 Idea 且利用 Maven 工具进行项目管理。下面介绍如何利用 Idea 构建一个基本的 Spring Maven 项目。

1.4.1 环境准备

（1）已安装 Java 17。本书所有章节代码都基于 Spring 6 构建。Spring 6 支持的最低 Java 版本为 Java 17，因此需要确保在计算机上已安装 Java 17，并配置环境变量 JAVA_HOME 为 Java 17 的安装路径。本书所使用的 Java 版本为 Java 17.0.4。

（2）安装 Idea 并配置 Maven。Idea 版本为 2023。本书后续项目都使用 Maven 管理，Idea 自带 Maven 插件，不需额外安装。每个 Maven 项目都有一个 pom.xml 文件，Maven 中的依赖包是基于 pom.xml 文件里面的三个标签<groupId>、<artifactId>、<version>来定位的。其中，groupId 为项目组 id，一般为公司域名的倒写。artifactId 为构件 id，一般为项目名。version 为项目的版本。例如要引入 Spring 的 webmvc 相关 Jar 包，版本为 6.0.3。可以在 pom.xml 文件中添加<dependencies>标签，在<dependencies>标签内部添加如下信息。

```
<dependency>
    <groupId>org.springframework</groupId>
    <artifactId>spring-webmvc</artifactId>
    <version>6.0.3</version>
</dependency>
```

除了<dependencies>标签外，pom.xml 文件里也提供了很多其他标签，在后续编程中如有出现，再进行介绍。

1.4.2 创建 Spring Maven 项目

下面展示如何利用 Idea 创建一个简单的 Spring Maven 项目。

创建 Spring Maven 项目

建立 Spring 开发环境步骤说明

1.4.3 认识注解

在 Spring 编程中注解被大量使用。注解并不是 Spring 独有的，早在 Java 5 中就引入了注解的概念。Spring 在 Java 的基础上又扩充了很多新的注解，以方便编程。使用注解首先要学会看懂注解，这里就必须要了解一个概念——元注解。元注解用于描述注解，元注解本身也是一种注解，但是它能够应用到其他注解上面，对之进行解释说明。

元注解分为@Retention、@Documented、@Target、@Inherited、@Repeatable 五种，下面做简单介绍。

1. @Retention

@Retention 用于描述注解的生命周期。@Retention 注解有三种取值，具体如表1-1所示。

表1-1 @Retention 注解

取值类型	描 述	使用方法
RetentionPolicy.SOURCE	表示被修饰的注解只在源代码中保留，编译器不保留	@Retention(RetentionPolicy.SOURCE)
RetentionPolicy.CLASS	表示被修饰的注解只在.class 文件中，JVM 不保留	@Retention(RetentionPolicy.CLASS)
RetentionPolicy.RUNTIME	表示被修饰的注解会被 JVM 保留，在运行时有效	@Retention(RetentionPolicy.RUNTIME)

2. @Documented

@Documented 用于描述注解可以被加入 Javadoc 工具文档中。

3. @Target

@Target 用于描述注解的使用范围，包括类、方法、变量等。@Target 取值如表1-2所示。

表1-2 @Target 注解

取值类型	描 述	使用方法
ElementType.ANNOTATION_TYPE	表示被修饰的注解只能应用在其他注解上	@Target(ElementType.ANNOTATION_TYPE)
ElementType.CONSTRUCTOR	表示被修饰的注解只能应用在构造函数上	@Target(ElementType.CONSTRUCTOR)
ElementType.FIELD	表示被修饰的注解只能应用在类属性上	@Target(ElementType.FIELD)
ElementType.LOCAL_VARIABLE	表示被修饰的注解只能应用在局部变量上	@Target(ElementType.LOCAL_VARIABLE)
ElementType.METHOD	表示被修饰的注解只能应用在方法上	@Target(ElementType.METHOD)
ElementType.PACKAGE	表示被修饰的注解只能应用在包上	@Target(ElementType.PACKAGE)
ElementType.PARAMETER	表示被修饰的注解只能应用在方法参数上	@Target(ElementType.PARAMETER)
ElementType.TYPE	表示被修饰的注解只能应用在类、接口、枚举等元素上	@Target(ElementType.TYPE)

4. @Inherited

被@Inherited 修饰的注解作用在某父类上，如果该父类的子类没有被任何其他注解修饰，那么这个子类就继承了父类的注解。具体使用方法如下。

```
/*定义 Father 注解,被@Inherited 修饰*/
@Inherited
public @interface Father {
}
/*定义 Person 父类,被 Father 注解修饰*/
@Father
class Person{}
/*定义 Person 的子类 man,也被 Father 注解修饰*/
class man extends Person{}
```

5. @Repeatable

@Repeatable 用于描述该注解可重复使用在同一类、方法和属性上。例如使用注解@PropertySources 加载多个配置文件，可以采用以下写法：

```
@PropertySources({@PropertySource(value="config1.xml"),
                  @PropertySource(value="config2.xml")})
public class Config{
}
```

这是因为@PropertySources 注解中有以下定义：

```
public @interface PropertySources {
    PropertySource[] value();
}
```

其中，PropertySource 也是一个注解，并被@Repeatable 修饰，修饰代码如下：

```
@Repeatable(PropertySources.class)
public @interface PropertySource {
}
```

因此，才能在@PropertySources 中多次重复使用@PropertySource 加载配置文件。

Spring 中的注解非常多，但它们共同被这五大元注解组合修饰。

1.4.4 基于注解方式的 Spring 编程

在 1.3.2 节中提到了我们可以通过 ApplicationContext 容器来管理 Bean。要获取具体的 Bean 有两种方式，分别是基于配置文件的方式和基于注解方式，其中基于注解方式因使用便捷成为项目开发中的主流。因此，这里重点介绍基于注解方式获取 Bean。

基于注解方式获取 Bean 要定义一个 Java 类作为 Spring 配置类并以@Configuration 注解修饰，在配置类中使用注解生成不同的 Bean。获取 Bean 时，首先获取 ApplicationContext 对象，然后调用 getBean 方法通过 id 或类型获取具体的 Bean。本书后续大部分 Spring 相关配置均基于注解方式实现。

下面介绍如何以注解方式实例化 ApplicationContext 容器。

在 src/main/java 目录下分别创建两个文件夹 bean 和 config,如图 1-3 所示。bean 文件夹用于存放 JavaBean,config 文件夹用于存放 Spring 配置类。

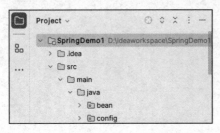

图 1-3　创建文件夹 bean 和 config

基于注解的 Spring 基本编程

在 config 文件夹下创建 SpringDemoConfig.java 文件作为 Spring 配置类,在 SpringDemoConfig.java 中输入以下代码。其中,@Configuration 用于将该类标识为 Spring 配置类,@ComponentScan("bean")用于设置生成 Bean 时自动扫描路径为 ClassPath 项目下的 bean 文件夹。

【SpringDemoConfig.java】

```
package config;
import org.springframework.context.annotation.ComponentScan;
import org.springframework.context.annotation.Configuration;
@Configuration
@ComponentScan("bean")
public class SpringDemoConfig {
}
```

在 bean 文件夹下创建类 Demo1.java。该类中提供一个 say 方法,该方法用于控制台输出 hello demo1。Demo1.java 中代码如下,其中,@Component 注解用于将 Demo1 这个类标识为一个 Spring 组件类的 Bean。

【Demo1.java】

```
package bean;
import org.springframework.stereotype.Component;
@Component
public class Demo1 {
    public void say(){
        System.out.println("hello demo1");
    }
}
```

在项目根目录 src/main/java 下创建一个测试类 BeanTest.java,在测试类中调用 Demo1 类中的 say 方法打印输出 hello demo1。传统方式需要编写以下代码,手动创建一个 Demo1 实例对象,然后调用 Demo1 对象的 say 方法。

```
Demo1 demo1=new Demo1();
demo1.say();
```

使用 Spring 后,Demo1 实例对象的创建交给 Spring 容器来做,创建完毕可从 Spring 容器中利用 getBean 方法获取 Demo1 实例对象并调用 say 方法输出内容。在 BeanTest.java 中输入以下代码,代码中通过 id 和类型两种不同方式获取实例化 Demo1 对象。

```
import bean.Demo1;
import config.SpringDemoConfig;
import org.springframework.beans.factory.BeanFactory;
import org.springframework.context.ApplicationContext;
import org.springframework.context.annotation.AnnotationConfigApplicationContext;
public class BeanTest {
    public static void main(String[] args){
        //实例化 ApplicationContext 容器
        ApplicationContext ctx=
        new AnnotationConfigApplicationContext(SpringDemoConfig.class);
        Demo1 deCtxByName=(Demo1)ctx.getBean("demo1");
        deCtxByName.say();
        Demo1 deCtxByType=ctx.getBean(Demo1.class);
        deCtxByType.say();
    }
}
```

运行 BeanTest.java 即可在控制台看到分别打印出 2 次 hello demo1,如图 1-4 所示。证明通过 Bean 的 id 和类型两种方法均可获取实例化 Bean 并调用方法输出结果。

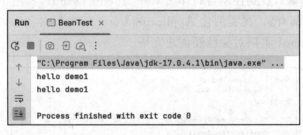

图 1-4 输出 hello demo1

修改 Demo1.java 文件,在其中添加一个无参数构造方法,在构造方法中添加打印语句。下面所示粗体代码为新增内容。

【Demo1.java】

```
package bean;
import org.springframework.stereotype.Component;
@Component
public class Demo1 {
    public Demo1(){
        System.out.println("Constructor Demo1");
    }
```

```
    public void say(){
        System.out.println("hello demo1");
    }
}
```

修改 BeanTest.java 文件，main 方法中只保留第一条语句，用于获取 ApplicationContext 对象。注释如下面黑体代码。

【BeanTest.java】

```
import bean.Demo1;
import config.SpringDemoConfig;
import org.springframework.beans.factory.BeanFactory;
import org.springframework.context.ApplicationContext;
import org.springframework.context.annotation.AnnotationConfigApplicationContext;
public class BeanTest {
    public static void main(String[] args){
        //实例化 ApplicationContext 容器
        ApplicationContext ctx=
        new AnnotationConfigApplicationContext(SpringDemoConfig.class);
        /* Demo1 deCtxByName=(Demo1)ctx.getBean("demo1");
        deCtxByName.say();
        Demo1 deCtxByType=ctx.getBean(Demo1.class);
        deCtxByType.say(); */
    }
}
```

运行 BeanTest.java，看到控制台打印出 Constructor Demo1，如图 1-5 所示。代码中即使不获取 Demo1 实例对象，在实例化 ApplicationContext 容器对象时，Demo1 的无参构造方法已经被调用，Demo1 实例对象已经创建完毕。

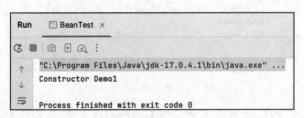

图 1-5　输出 Constructor Demo1

任务 1.5　了解 Spring 控制反转（IOC）

在传统编程中，大多数应用程序都是由多个类通过彼此的合作来实现复杂的业务逻辑，当需要使用类的对象时，一般会直接在程序中直接创建，对象和程序本身形成了硬编码关系，耦合度太高。如果能够将创建对象的权限转移给第三方容器实现，需要时再通过第三方容器获取对象的引用，这样能够降低代码的耦合度，有利于代码维护。这正是 IOC 产生的原因。

1.5.1 初识 IOC

控制反转 IOC,是软件设计中一种思想。控制反转能够使服务变更的控制权从服务提供者转向服务调用者,有利于降低代码耦合度,减轻硬编码的问题。下面介绍 IOC 的思想。

这里模拟一个类似 MVC 三层架构方式的服务调用,服务功能为获取用户数据。在 SpringDemo1 项目中新建 dao 和 service 两个文件夹,分别创建 UserDao.java、UserDaoImpl1.java、UserDaoImpl2.java、UserService.java、UserServiceImpl.java 这 5 个 Java 类,结构如图 1-6 所示。其中 UserService 为对外服务提供的接口,供用户调用。UserServiceImpl 为服务实现类继承 UserService 接口。UserDao 为 Dao 层接口,UserDaoImpl1 和 UserDaoImpl2 为 Dao 层的不同实现类均继承 UserDao 接口,UserDaoImpl1 和 UserDaoImpl2 分别采用 A 和 B 两种方式获取用户数据。

图 1-6 结构图

在 UserDao.java、UserDaoImpl1.java、UserDaoImpl2.java、UserService.java、UserServiceImpl.java 这 5 个 Java 类中写入以下代码:

【UserDao.java】

```
package dao;
public interface UserDao {
    void getUser();
}
```

【UserDaoImpl1.java】

```
package dao;
public class UserDaoImpl1 implements UserDao {
    @Override
    public void getUser() {
        System.out.println("采用A方式获取用户数据");
    }
}
```

【UserDaoImpl2.java】

```java
package dao;
public class UserDaoImpl2 implements UserDao {
    @Override
    public void getUser() {
        System.out.println("采用B方式获取用户数据");
    }
}
```

【UserService.java】

```java
package service;
public interface UserService {
    void getUser();
}
```

【UserServiceImpl.java】

```java
package service;
import dao.UserDao;
import dao.UserDaoImpl1;
public class UserServiceImpl implements UserService {
    private UserDao userdao=new UserDaoImpl1();
    @Override
    public void getUser() {
        userdao.getUser();
    }
}
```

假设用户需要用 A 方式获取数据，在 src/main/java 目录下新建 ServiceTest.java 文件作为服务测试类模拟 Controller 层，里面提供 main 方法对 UserService 提供的服务进行调用。

【ServiceTest.java】

```java
import service.UserService;
import service.UserServiceImpl;
public class ServiceTest {
    public static void main(String[] args){
        UserService userService=new UserServiceImpl();
        userService.getUser();
    }
}
```

ServiceTest.java 运行结果如下：

```
采用A方式获取用户数据
```

如果用户有新需求，需用 B 方式获取数据，那么业务层实现类 UserServiceImpl.java，需要做相应改动，即将：

```
private UserDao userdao=new UserDaoImpl1();
```

改成：

```
private UserDao userdao=new UserDaoImpl2();
```

这种方式下，服务变更的控制权在服务提供者手中，需要程序员直接对业务层代码进行修改，用户变更一个新需求，业务层代码就需要相应改变。如果业务层代码涉及很多 Java 文件，改动代码量巨大，这是很不科学的一种代码架构。实际上，这种改动应该由调用者 ServiceTest.java 触发，服务提供者只需把各种各样的服务提供出来。用户根据自己需要选择对应的服务方式。就像在 Idea 中更换插件一样，Idea 在商城中提供了很多插件，我们需要更换什么插件就把原有插件删除，再安装新的插件，而不是由 Idea 的维护人员帮我们更换插件。

为了解决这个问题，将代码做如下粗体字所示修改。修改 UserServiceImpl.java 代码，在其中定义私有变量 userdao 且不赋值。添加 setUserdao 方法，为 userdao 变量注入不同实例对象。同时 UserService 接口中添加 setUserdao 抽象方法。

【UserServiceImpl.java】

```java
package service;
import dao.UserDao;
import dao.UserDaoImpl1;
public class UserServiceImpl implements UserService {
    //private UserDao userdao=new UserDaoImpl1();
    private UserDao userdao;
    @Override
    public void getUser() {
        userdao.getUser();
    }
    //set 方法设置注入
    public void setUserdao(UserDao userdao) {
        this.userdao = userdao;
    }
}
```

【UserService.java】

```java
package service;
import dao.UserDao;
public interface UserService {
    void getUser();
    void setUserdao(UserDao userdao);
}
```

此时，业务层代码不再有 new UserDaoImpl1()，业务层代码交出了服务控制权。调用者可以根据需要选择创建 UserDaoImpl1 还是 UserDaoImpl2 对象，利用 setUserdao 方法把创建的对象注入 UserServiceImpl 的 userdao 中。UserServiceImpl 由主动创建对象变为被动接收对象，实现服务控制权的反转。ServiceTest.java 代码中修改如下粗体代码。

【ServiceTest.java】

```java
import dao.UserDaoImpl2;
import service.UserService;
import service.UserServiceImpl;
public class ServiceTest {
    public static void main(String[] args){
        UserService userService=new UserServiceImpl();
        userService.setUserdao(new UserDaoImpl2());
        userService.getUser();
    }
}
```

此时 ServiceTest.java 运行结果如下：

```
采用 B 方式获取用户数据
```

1.5.2 依赖注入

控制反转的实现方式有多种，依赖注入只是实现控制反转的一种方式。在 1.5.1 小节中利用 setUserdao 方法为 userdao 变量注入值这种方式即为依赖注入，其中，set 方法只是实现依赖注入的方法之一，后续还会介绍到其他方法。在使用 Spring 框架之后，UserDaoImpl2 对象实例不再由调用者创建，而是由 Spring 容器来创建，这样服务控制权由调用者又转移到了 Spring 的容器。

这时，如果需要将用户数据获取方式从 A 改成 B，我们只需将 userdao 变量注入值由 UserDaoImpl1 对象换成 UserDaoImpl2 对象的变化告诉 Spring 的容器，由 Spring 的容器帮我们自动变更。如何告诉 Spring 的容器呢？有两种方式：基于 XML 文件的配置和基于 Config 注解类的配置。基于 XML 文件的配置比较烦琐，这里只介绍基于 Config 注解类的配置。

对原有的 SpringDemoConfig.java 类按如下粗体代码修改。

【SpringDemoConfig.java】

```java
package config;
import dao.UserDao;
import dao.UserDaoImpl1;
import dao.UserDaoImpl2;
import org.springframework.context.annotation.Bean;
import org.springframework.context.annotation.ComponentScan;
import org.springframework.context.annotation.Configuration;
import service.UserServiceImpl;
@Configuration
```

```
@ComponentScan("bean")
public class SpringDemoConfig {
    @Bean
    public UserDaoImpl1 userDaoImpl1(){
        return new UserDaoImpl1();
    }
    @Bean
    public UserDaoImpl2 userDaoImpl2(){
        return new UserDaoImpl2();
    }
    @Bean
    public UserService userServiceImpl(UserDaoImpl1 userDaoImpl){
        UserService userServiceImpl=new UserServiceImpl();
        //调用setUserdao方法进行set设值注入
        userServiceImpl.setUserdao(userDaoImpl);
        return userServiceImpl;
    }
}
```

上述代码依次在配置类中利用@Bean注解创建名为userDaoImpl1和userDaoImpl2的Bean，然后利用@Bean注解创建名为userServiceImpl的Bean，方法输入参数类型为UserDaoImpl1类型。内部创建userServiceImpl类型的Bean并调用UserServiceImpl类的setUserdao方法将UserDaoImpl1类型的对象参数设值注入userServiceImpl类型的Bean中，实现依赖注入。

对原有测试类ServiceTest.java类按如下粗体代码修改。

【ServiceTest.java】

```
import dao.UserDaoImpl2;
import org.springframework.context.ApplicationContext;
import org.springframework.context.annotation.AnnotationConfigApplicationContext;
import service.UserService;
import service.UserServiceImpl;
public class ServiceTest {
    public static void main(String[] args){
        //注释原有代码
        /* UserService userService=new UserServiceImpl();
        userService.setUserdao(new UserDaoImpl2());
        userService.getUser(); */
        //新增代码
        ApplicationContext ctx=
        new AnnotationConfigApplicationContext(SpringDemoConfig.class);
        UserService userService=(UserService)ctx.getBean("userServiceImpl");
        userService.getUser();
    }
}
```

ServiceTest.java运行结果如下：配置类中注入UserDaoImpl1对象。

采用 A 方式获取用户数据

此时如果需要将用户数据获取方式从 A 改成 B,只需改动 Spring 配置类,将其中 userServiceImpl 方法的输入参数类型从 UserDaoImpl1 改成 UserDaoImpl2,这时 setUserdao 方法注入的对象类型就变成 UserDaoImpl2。原有业务层 UserServiceImpl.java 代码不需改动,测试类 ServiceTest.java 代码也不需改动,服务控制权完全交给了 Spring。

任务 1.6　综合案例: 获取 Spring 中 Bean 的相关信息

基于前面已介绍的 Spring 相关知识,这里以一个综合案例演示 Spring 容器和 Spring IOC 编程的综合应用,案例实现功能为编程输出 Spring 容器中所有 Bean 的相关信息。

1.6.1　案例任务

任务内容:分别编写 Dao 层和 Service 层来实现类和接口,并添加相应方法实现遍历 Spring 容器中所有的 Bean,输出所有 Bean 的名字和类型。编写测试类调用方法并在控制台打印结果。

1.6.2　任务分析

该任务主要利用依赖注入思想实现编程,需新建 Dao 层和 Service 层的 Java 类,将 Dao 层对象注入 Service 层,测试类获取 Service 层对象并调用方法来打印结果。任务实现思路如下。

(1)新建 Dao 层接口和实现类,在实现类中编写方法,遍历 Spring 容器中所有的 Bean,输出所有 Bean 的名字和类型。

(2)新建 Service 层接口和实现类,将 Dao 层注入 Service 层实现类中。

(3)在 Spring 配置类中配置 Dao 层和 Service 层的 Bean。

(4)编写测试类,从 Spring 容器中获取 Servcie 层的 Bean 并调用方法来打印结果。

1.6.3　任务实施

在 SpringDemo1 项目的 dao 目录下新建 dao 层接口和实现类,SpringBeanDao.java 为 dao 层接口,SpringBeanDaoImpl.java 为 dao 层实现类继承的 SpringBeanDao 接口。SpringBeanDaoImpl 内部实现的 getSpringBean 方法用于遍历输出 Spring 容器中所有 Bean 的名字和类型。

【SpringBeanDao.java】

```
package dao;
import org.springframework.context.ApplicationContext;
public interface SpringBeanDao {
```

```
    void getSpringBean(ApplicationContext ctx);
}
```

【SpringBeanDaoImpl.java】

```
package dao;
import org.springframework.context.ApplicationContext;
import java.util.Arrays;
public class SpringBeanDaoImpl implements SpringBeanDao{
    @Override
    public void getSpringBean(ApplicationContext ctx) {
        //获取Spring容器中所有的Bean的id
        String [] arr=ctx.getBeanDefinitionNames();
        //利用Java流式编程输出Bean的id和类型
        Arrays.stream(arr).forEach(str->System.out.println("Bean的id为:"+str
        +",Bean的类型为:"+ctx.getBean(str).getClass()));
    }
}
```

getSpringBean方法使用ApplicationContext类型参数,利用ApplicationContext对象中的getBeanDefinitionNames方法获取Spring容器中所有Bean的id。编程时,当通过Bean的id获取Bean对象失败时,可通过调用getBeanDefinitionNames方法显示Spring容器中所有Bean的id,判断该Bean是否存在,从而快速定位错误。

在SpringDemo1项目的service目录下新建service层的接口和实现类。SpringBeanService.java为service层接口,SpringBeanServiceImpl.java为service层实现类并继承SpringBeanService接口。SpringBeanServiceImpl类内部实现的getSpringBeanService方法用于调用SpringBeanDaoImpl中的getSpringBean方法来遍历输出Spring容器中所有Bean的名字和类型,同时添加setSpringBeanDao方法注入SpringBeanDao对象。

【SpringBeanService.java】

```
package service;
import dao.SpringBeanDao;
import org.springframework.context.ApplicationContext;
public interface SpringBeanService {
    void getSpringBeanService(ApplicationContext ctx);
    void setSpringBeanDao(SpringBeanDao springBeanDao);
}
```

【SpringBeanServiceImpl.java】

```
package service;
import dao.SpringBeanDao;
import org.springframework.context.ApplicationContext;
public class SpringBeanServiceImpl implements GetSpringBeanService {
    private SpringBeanDao springBeanDao;
    //set方法依赖注入SpringBeanDao对象给springBeanDao私有变量
    public void setSpringBeanDao(SpringBeanDao springBeanDao) {
```

```
        this.springBeanDao = springBeanDao;
    }
    @Override
    public void getSpringBeanService(ApplicationContext ctx) {
        springBeanDao.getSpringBean(ctx);
    }
}
```

在 SpringDemoConfig.java 配置类中配置 SpringBeanDao 和 SpringBeanService 的 Bean，编写如下粗体代码。

【SpringDemoConfig.java】

```
import dao.*;
import org.springframework.context.annotation.Bean;
import org.springframework.context.annotation.ComponentScan;
import org.springframework.context.annotation.Configuration;
import service.SpringBeanService;
import service.SpringBeanServiceImpl;
import service.UserService;
import service.UserServiceImpl;
@Configuration
@ComponentScan("bean")
public class SpringDemoConfig {
    @Bean
    public UserDaoImpl1 userDaoImpl1(){
        return new UserDaoImpl1();
    }
    @Bean
    public UserDaoImpl2 userDaoImpl2(){
        return new UserDaoImpl2();
    }
    @Bean
    public UserService userServiceImpl(UserDaoImpl1 userDaoImpl){
        UserService userServiceImpl=new UserServiceImpl();
        //调用 setUserdao 方法进行 set 设值注入
        userServiceImpl.setUserdao(userDaoImpl);
        return  userServiceImpl;
    }
    //配置 SpringBeanDao 和 SpringBeanService 的 Bean
    @Bean
    public SpringBeanDao getSpringBeanDao(){return new SpringBeanDaoImpl();}
    @Bean
    public SpringBeanService getSpringBeanService(SpringBeanDao springBeanDao){
        SpringBeanService springBeanServiceImpl=new SpringBeanServiceImpl();
        springBeanServiceImpl.setSpringBeanDao(springBeanDao);
        return springBeanServiceImpl;
    }
}
```

在 src/main/java 目录下创建测试类 AllBeanTest.java，在类中编写以下代码。

【AllBeanTest.java】

```java
import config.SpringDemoConfig;
import org.springframework.context.ApplicationContext;
import org.springframework.context.annotation.AnnotationConfigApplicationContext;
import service.SpringBeanService;
public class AllBeanTest {
    public static void main(String[] args){
        //获取 ApplicationContext 容器
        ApplicationContext ctx=
        new AnnotationConfigApplicationContext(SpringDemoConfig.class);
        //通过 Bean 类型获取 SpringBeanService Bean 对象
        SpringBeanService springBeanService=
        ctx.getBean(SpringBeanService.class);
        //调用方法输出 Spring 容器中所有 Bean 的 id 和类型
        SpringBeanService.getSpringBeanService(ctx);
    }
}
```

代码运行结果如图 1-7 所示。从图 1-7 中可以看出 Spring 框架运行时自身会默认创建一些内部的 Bean（图中前 4 个 Bean）。随后是我们在 SpringDemoConfig.java 文件中配置的一些 Bean。包括以下几种。

```
"C:\Program Files\Java\jdk-17.0.4.1\bin\java.exe" ...
Bean的id为: org.springframework.context.annotation.internalConfigurationAnnotationProcessor,Bean的类型为:
Bean的id为: org.springframework.context.annotation.internalAutowiredAnnotationProcessor,Bean的类型为: class
Bean的id为: org.springframework.context.event.internalEventListenerProcessor,Bean的类型为: class org.sprin
Bean的id为: org.springframework.context.event.internalEventListenerFactory,Bean的类型为: class org.springfr
Bean的id为: springDemoConfig,Bean的类型为: class config.SpringDemoConfig$$SpringCGLIB$$0
Bean的id为: demo1,Bean的类型为: class bean.Demo1
Bean的id为: userDaoImpl1,Bean的类型为: class dao.UserDaoImpl1
Bean的id为: userDaoImpl2,Bean的类型为: class dao.UserDaoImpl2
Bean的id为: userServiceImpl,Bean的类型为: class service.UserServiceImpl
Bean的id为: getSpringBeanDao,Bean的类型为: class dao.GetSpringBeanDaoImpl
Bean的id为: getSpringBeanService,Bean的类型为: class service.GetSpringBeanServiceImpl
```

图 1-7　AllBeanTest.java 程序的运行结果

ConfigurationClassPostProcessor：Spring 处理配置类的后置处理器，用于对带有 @Configuration 注解的 SpringDemoConfig 类进行处理并生成 Bean。

AutowiredAnnotationBeanPostProcessor：用于对 Bean 实现自动装配，如 Dao 层对象利用 set 方法注入 Service 层对象。

DefaultEventListenerFactory：Spring 启动时默认的 Bean 工厂。

EventListenerMethodProcessor：对 DefaultEventListenerFactory 提供支持。用于 Bean 的发现与注册，如在 SpringDemoConfig.java 文件中用 @Bean 注解标识的一系列 Bean 都将被发现并注册进 Spring 容器中。

小　　结

　　Spring Boot 是 Spring 框架的一部分，学习 Spring Boot 之前首先要了解 Spring 框架。本项目详细介绍了 Spring 框架的相关知识，包括 Spring 简介、Spring 的作用、Spring 的发展历程、Spring 容器的概念、IOC 控制反转、依赖注入以及 Spring 的初步编程。最后以一个综合案例演示如何编程获取 Spring 容器对象以及 Bean 的相关信息，使读者对 Spring 框架有初步的认识。

课后练习：获取 Spring Bean 对象相关信息并过滤

　　新建一个 Spring Maven 项目，再创建 Dao 层和 Service 层。在 Dao 层中新建 Java 类并添加 2 个方法，方法功能如下。

　　方法 1：获取 Spring 容器中 Bean 对象的数量。

　　方法 2：过滤并输出 Spring 容器中所有的 Bean。只输出所有自定义 Bean，即 Bean 的 id 包含 springframework 关键字的不输出。

　　然后将 Dao 层注入 Service 层的 Java 类中，并编写一个测试类调用方法 1 和方法 2，在控制台输出运行结果。

项目 2　认识 Spring 中的 Bean

目前 Spring 应用程序大多使用注解开发，其中最重要的就是如何利用注解实现 Bean 的创建和管理。Spring 中的 Bean 是基于 Java 实体类创建的。按照实体类有无成员变量划分，Java 实体类可以分为两类：一类是类内部没有变量属性或属性为常量，在创建 Bean 时无须外部赋值，可直接调用类默认无参构造方法；另一类是类内部拥有变量属性的，在创建 Bean 时需外部传入数据对属性赋值。本项目将对 Spring 中的 Bean 做详细介绍，包括无变量属性和有变量属性 Bean 的创建、Bean 的作用域、Bean 的生命周期等内容。

任务 2.1　基于注解创建无变量属性 Bean

Spring 基于注解方式创建无变量属性 Bean 有以下几种方法。
(1) 在 Spring 配置类中通过@Bean 标识方法创建 Bean。
(2) 在 Spring 配置类中添加@ComponentScan，配置自动扫描路径，并将该路径下相关 Java 类利用注解进行标识。如在项目 1 的 1.4.4 小节中，在 Spring 配置类中配置@ComponentScan("bean")，同时用@Component 注解 Demo1 类。
(3) 在 Spring 配置类中通过@Import 创建 Bean。
(4) 通过 FactoryBean 工厂创建 Bean。
下面将详细介绍这 4 种创建 Bean 的方法。

2.1.1　通过@Bean 标识方法创建 Bean

@Bean 注解一般应用在方法上，也可以应用在其他注解上。被@Bean 注解标注的方法会生成一个由 Spring 容器管理的 Bean。其中，方法的返回值类型为 Bean 的类型，方法名为 Bean 的 id。@Bean 注解一般和@Configuration 注解一同使用，用于在 Spring 配置类中配置并生成 Bean。

在使用时，@Bean 注解内部还有 5 个属性可以设置，这 5 个属性含义如表 2-1 所示。

表 2-1　@Bean 注解内部属性

属　性	作　用	使 用 方 法
name	用于指定 Bean 的 id。如果不指定，采用方法名为 Bean 的 id；如果指定多个值，第一个值为名称为 Bean 的 id，其余值为别名	@Bean(name="u")，其中 u 为 Bean 的 id。Bean id 是唯一的，而 Bean 别名可以有多个

续表

属性	作用	使用方法
value	作用和 name 类似	@Bean(value={"u1","u2"}),其中 Bean 的 id 为 u1,别名为 u2,在配置中,name 和 value 两个属性使用其中一个即可
autowireCandidate	用于指定该 Bean 对象能否自动注入其他 Bean 中,默认为 true	@Bean(autowireCandidate=false),禁止当前 Bean 注入其他 Bean 中
initMethod	用于指定 Bean 的初始化方法,默认为空,表示不调用初始化方法	@Bean(initMethod="initMethod"),使用时直接指定方法名称,此处指定 Bean 初始化方法为 initMethod
destroyMethod	用于指定 Bean 的销毁方法,默认为 inferred,destroyMethod 设置只对单例 Bean 销毁有效	@Bean(destroyMethod="destroyMethod"),使用时直接指定方法名称,此处指定 Bean 销毁方法为 destroyMethod

下面演示这 5 个属性的作用。新建一个 Spring Maven 项目 SpringDemo2,在该项目的根目录下新建 bean 文件夹,在 bean 文件夹内新建 Bean1.java 文件,内部定义初始化方法 initMethod 和销毁方法 destoryMethod。

【Bean1.java】

```java
package bean;
public class Bean1 {
    //Bean1 的初始化方法
    public void initMethod(){
        System.out.println("initMethod Bean1");
    }
    //Bean1 的销毁方法
    public void destoryMethod(){
        System.out.println("destory Method Bean1");
    }
}
```

在根目录下新建 config 文件夹,在 config 文件夹内新建 SpringDemoConfig.java 文件,在文件内分别配置 3 个不同名字 Bean1 类型的 Bean,代码如下。

【SpringDemoConfig.java】

```java
package config;
import bean.Bean1;
import bean.DiToBean1;
import org.springframework.context.annotation.Bean;
import org.springframework.context.annotation.ComponentScan;
import org.springframework.context.annotation.Configuration;
@Configuration
public class SpringDemoConfig {
    //Bean 的 id 为 getBean1
    @Bean
    public Bean1 getBean1(){
```

```
        return new Bean1();
    }
    //Bean 名称为 u1,u2 为别名
    @Bean(name = {"u1","u2"})
    public Bean1 getBean2(){
        return new Bean1();
    }
    //Bean 名称为 u3,u4 为别名,创建对象将调用 initMethod 方法
    //销毁对象将调用 destoryMethod 方法
    @Bean(name = {"u3","u4"},initMethod ="initMethod",
        destroyMethod = "destoryMethod")
    public Bean1 getBean3(){
        return new Bean1();
    }
}
```

创建测试类 BeanTest.java,分别获取配置类中配置的 3 个 Bean,代码如下。

【BeanTest.java】

```
import bean.Bean1;
import config.SpringDemoConfig;
import org.springframework.context.ApplicationContext;
import org.springframework.context.annotation.AnnotationConfigApplicationContext;
public class BeanTest {
    public static void main(String [] args){
        AnnotationConfigApplicationContext ctx=
        new AnnotationConfigApplicationContext(SpringDemoConfig.class);
        Bean1 bean1=(Bean1)ctx.getBean("getBean1");
        System.out.println("成功获取名为 getBean1 对象");
        Bean1 bean2=(Bean1)ctx.getBean("u1");
        System.out.println("成功获取名为 u1 对象");
        Bean1 bean3=(Bean1)ctx.getBean("u3");
        System.out.println("成功获取名为 u3 的对象,并调用 u3 对象初始化方法");
        ctx.close();
        System.out.println("成功调用 u3 对象销毁方法销毁对象 u3");
    }
}
```

由于在容器停止时 Spring 才会调用 destroyMethod 方法销毁 Bean。而停止容器的 close 方法只存在于 ApplicationContext 子类的 AnnotationConfigApplicationContext 对象中。因此,这里 ctx 变量类型为 AnnotationConfigApplicationContext,而不是 ApplicationContext。

Spring 默认创建的 Bean 是单例的,只有单例 Bean 在销毁时才会调用 destroyMethod 方法,非单例 Bean 销毁不会调用该方法。

BeanTest.java 程序的运行结果如图 2-1 所示。从图中打印顺序可以看出,所有的 Bean 都被创建,Bean 创建顺序按照 Spring 配置类代码顺序执行。因为名为 u3 的 Bean 配置了 initMethod 方法和 destroyMethod 方法,因此,在名为 u3 的 Bean 创建时首先执行 initMethod 方法来打印 initMethod Bean1。在执行 ctx.close()语句时,又调用了该 Bean 的 destroyMethod 方

法销毁对象并打印出 destoryMethod Bean1。

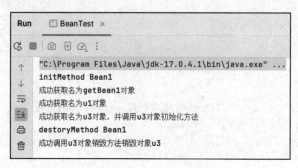

图 2-1 BeanTest.java 程序的运行结果

下面演示 autowireCandidate 属性的作用。在 bean 文件夹下新建 DiToBean1.java，在 DiToBean1 类中采用 set 方法注入 Bean1 类型的对象。

【DiToBean1.java】

```
package bean;
public class DiToBean1 {
    private Bean1 bean1;
    public void setBean1(Bean1 bean1) {
        this.bean1 = bean1;
    }
}
```

修改配置类 SpringDemoConfig.java，修改后的代码如下，其中，第一个 Bean 上添加了 autowireCandidate 属性并赋值为 false，使该 Bean 不能注入其他 Bean，同时配置生成 DiToBean1 类型的 Bean。

【SpringDemoConfig.java】

```
package config;
import bean.Bean1;
import bean.DiToBean1;
import org.springframework.context.annotation.Bean;
import org.springframework.context.annotation.ComponentScan;
import org.springframework.context.annotation.Configuration;
@Configuration
public class SpringDemoConfig {
    //Bean 的 id 为 getBean1,添加 autowireCandidate 并设为 false,使该 Bean 不可注入
      其他 Bean
    @Bean(autowireCandidate = false)
    public Bean1 getBean1(){
        return new Bean1();
    }
    @Bean
    public DiToBean1 getDiToBean1(Bean1 bean1){
        DiToBean1 diToBean1=new DiToBean1();
        diToBean1.setBean1(bean1);
```

```
        return diToBean1;
    }
}
```

修改测试类 BeanTest.java，修改的代码如下。看程序是否能成功获取 DiToBean1 类型的 Bean。

【BeanTest.java】

```
import bean.DiToBean1;
import config.SpringDemoConfig;
import org.springframework.context.ApplicationContext;
import org.springframework.context.annotation.AnnotationConfigApplicationContext;
public class BeanTest {
    public static void main(String [] args){
        AnnotationConfigApplicationContext ctx=
            new AnnotationConfigApplicationContext(SpringDemoConfig.class);
        DiToBean1 diToBean1=(DiToBean1)ctx.getBean("getDiToBean1");
    }
}
```

运行结果中对错误的提示信息如图 2-2 所示，从图 2-2 中可以看出 autowireCandidate 属性生效，报错信息为创建名为 getDiToBean1 的 Bean 失败，没有合适的 Bean1 类型的 Bean 可用。如果把 autowireCandidate 设置为 true，则 Bean1 对象能够成功注入 DiToBean1。

> Error creating bean with name 'getDiToBean1' defined in config.SpringDemoConfig: Unsatisfied dependency expressed through method 'getDiToBean1' parameter 0: No qualifying bean of type 'bean.Bean1' available: expected at least 1 bean which qualifies as autowire candidate. Dependency annotations: {}

图 2-2 注入 Bean1 失败时运行结果对错误的提示信息

2.1.2 通过@ComponentScan 自动扫描方式创建 Bean

在 2.1.1 小节中通过@Bean 标识方法创建 Bean，但是在实际开发过程中如果有很多 Bean 需要生成，逐个手动在 Spring 配置类里面使用@Bean 注解配置是很不方便的，因此 Spring 提供了@ComponentScan 注解，允许开发人员通过自动扫描方式创建 Bean。@ComponentScan 注解多用在配置类上，允许开发人员自定义扫描规则，然后根据定义的扫描规则找出具有特定注解标识的类，并自动装配到 Spring 的容器中。这种创建 Bean 的方式在开发大型项目时经常用到。

在 1.4.4 小节中，我们曾编写了以下代码，利用@ComponentScan 自动扫描 bean 目录下被@Component 标识的 Java 类来创建 Bean。

```
@Configuration
@ComponentScan("bean")                              //设置自动扫描范围为 bean 目录
public class SpringDemoConfig {
    //此处省略代码
```

```
}
@Component                              //标识 Demo1 为一个组件类 Bean
public class Demo1 {
    //此处省略代码
}
```

@ComponentScan 本身并不能创建 Bean,它只是一个扫描工具,必须配合其他注解使用,如 @Component。@Component 注解将 Demo1 类标识一个组件类 Bean,在 @ComponentScan 自动扫描时,扫描到 Demo1 类上标记有 @Component 注解,就将 Demo1 类加载到 Spring 容器并实例化。

除了 @Component 外,Spring 还提供了其他注解,用于标记不同种类的 Bean,这些注解和 @ComponentScan 配合使用,使创建 Bean 的过程更加方便。

在 Web 开发中一般采用 MVC 三层架构,每层都需要创建很多 Java 对象。Spring 针对三层架构都提供了相应的注解以方便生成 Bean,如 @Controller、@Service、@Repository 等。表 2-2 列出了 Web 开发中常用 Bean 的注解。

表 2-2 Web 开发中常用 Bean 的注解

注解名	作用
@Component	@Component 用于将 Java 类标识成通用的组件类 Bean
@Controller	@Controller 用于将 Java 类标识成控制层组件类的 Bean,对应于 MVC 三层架构的控制层
@Service	@Controller 用于将 Java 类标识成业务层组件类的 Bean,对应于 MVC 三层架构的业务层
@Repository	@Controller 用于将 Java 类标识成持久层组件类的 Bean,对应于 MVC 三层架构的持久层

注意:使用 @Component、@Controller、@Service、@Repository 生成的 Bean 名字默认为该 Java 类首字母小写的驼峰命名法。@Component、@Controller、@Service、@Repository 标记的 Bean 在使用上没有任何区别,只是开发人员根据 MVC 三层结构编程规范对 Bean 进行不同的分层。如果把 Demo1 类上的注解从 @Component 换成 @Controller。

```
@Controller                             //标识 Demo1 为一个控制层组件类 Bean
public class Demo1 {
    //此处省略代码
}
```

该 Bean 对象在 BeanTest.java 中一样能正常创建和使用,如图 2-3 所示。

2.1.3 通过 @Import 创建 Bean

@Import 注解只能用在 Java 类上,用于实例化 Bean 并加入 Spring 容器中。@Import 注解被大量运用于 Spring Boot 框架中。@Import 注解提供以下三种用法创建 Bean。

(1) 将要创建 Bean 的 Java 类直接在 Spring 配置类中应用 @Import 注解并创建 Bean。

(2) 创建一个实现了 ImportSelector 接口的注册类,在 Spring 配置类中应用 @Import

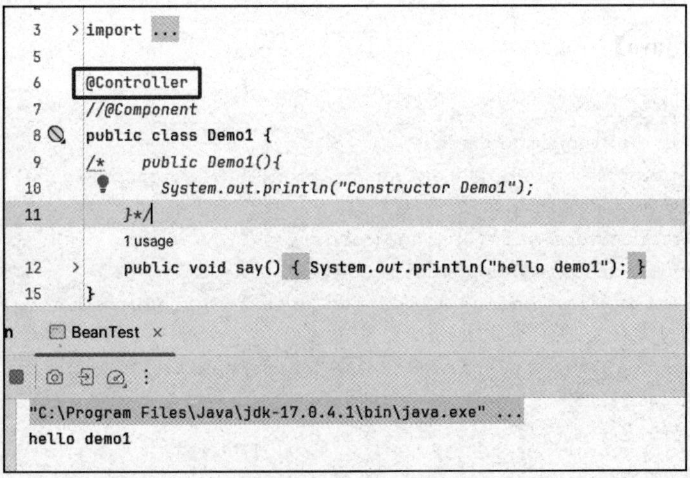

```
3    > import ...
5
6      @Controller
7      //@Component
8      public class Demo1 {
9          /*   public Demo1(){
10              System.out.println("Constructor Demo1");
11          }*/
           1 usage
12      >   public void say() { System.out.println("hello demo1"); }
15      }
```

```
BeanTest ×
"C:\Program Files\Java\jdk-17.0.4.1\bin\java.exe" ...
hello demo1
```

图 2-3 @Component 替换成@Controller

注解并创建 Bean。

(3) 创建一个继承了 ImportBeanDefinitionRegistrar 接口的注册类，在 Spring 配置类中应用@Import 注解并创建 Bean。

下面对三种方式做举例介绍。

1. 将要创建 Bean 的 Java 类直接在 Spring 配置类中应用@Import 注解并创建 Bean

此方法生成 Bean 的 id 是类的全路径，即类的路径.类名。

下面举例说明该方法的具体应用。在 Bean 目录下新建 Bean2.java，在 Bean2 类的构造方法中添加打印语句，同时在类中添加 say 方法。

【Bean2.java】

```
package bean;
public class Bean2 {
    public Bean2(){
        System.out.println("实例化 Bean2 对象");
    }
    public void say(){
        System.out.println("调用 Bean2 say 方法");
    }
}
```

修改 SpringDemoConfig.java 配置文件，在类上添加注解：

```
@Import(Bean2.class)
```

如需要添加多个类，则格式如下：

```
@Import({Bean2.class,Bean3.class})
```

此时生成的 Bean id 为 bean.Bean2。

新建测试类 ImportTest.java，在其中通过 id 为 bean.Bean2 获取 Bean2 对象并调用 say

方法。

【ImportTest.java】

```
import bean.Bean2;
import config.SpringDemoConfig;
import org.springframework.context.annotation.AnnotationConfigApplicationContext;
public class ImportTest {
    public static void main(String[] args){
        AnnotationConfigApplicationContext ctx=
        new AnnotationConfigApplicationContext(SpringDemoConfig.class);
        //通过类路径.类名获取 Bean2 对象
        Bean2 bean2=(Bean2)ctx.getBean("bean.Bean2");
        bean2.say();
    }
}
```

运行结果如图 2-4 所示,从图 2-4 中可以看到调用 Bean2 的构造方法创建对象,say 方法也成功调用。

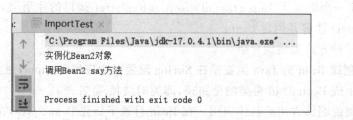

图 2-4 ImportTest.java 程序的运行结果 1

2. 创建一个实现了 ImportSelector 接口的注册类,并在 Spring 配置类中应用@Import 注解并创建 Bean

此方法要求注册类继承 ImportSelector 接口,并重写 selectImports 方法,该方法返回一个字符串数组的对象,数组里面的元素为注册 Bean 的 id,Bean 的 id 一般需给定全路径,即类路径.类名。所有数组里面的类都会注入 Spring 容器中。下面还是以创建 Bean2 对象举例说明此方法的具体应用。

在 bean 文件夹中新建 MyBeanImportSelector.java,该类继承 ImportSelector 接口,并重写 selectImports 方法。selectImports 方法返回一个字符串数组,数组中使用如下语句,

```
Bean2.class.getName()
```

获取 Bean2 类的全路径 bean.Bean2 作为生成 Bean 的 id。MyBeanImportSelector.java 代码如下。

【MyBeanImportSelector.java】

```
package bean;
import org.springframework.context.annotation.ImportSelector;
import org.springframework.core.type.AnnotationMetadata;
public class MyBeanImportSelector implements ImportSelector {
```

```
    @Override
    public String[] selectImports(AnnotationMetadata importingClassMetadata) {
        return new String[]{Bean2.class.getName()};
    }
}
```

修改 SpringDemoConfig.java 配置文件如下：

```
/*@Import(Bean2.class)*/
@Import(MyBeanImportSelector.class)
```

注释掉 Bean2 类，导入 MyBeanImportSelector 类。运行 ImportTest.java，也能看到如图 2-4 所示的运行结果。

3. 创建一个继承了 ImportBeanDefinitionRegistrar 接口的注册类，并在 Spring 配置类中应用 @Import 注解并创建 Bean

此方法要求注册类继承 ImportBeanDefinitionRegistrar 接口，同时重写 registerBeanDefinitions 方法，在 registerBeanDefinitions 方法中利用 Bean 注册器对象 BeanDefinitionRegistry 往 Spring 容器中注册 Bean。下面还是以创建 Bean2 对象为例说明此方法的具体应用。

在 Bean 文件夹下新建注册类 MyBeanImportRegistry.java，该注册类继承 ImportBeanDefinitionRegistrar 接口并重写 registerBeanDefinitions 方法。设置 Bean2 类型的 Bean id 为 MyBean2 并注册 Spring 容器。MyBeanImportRegistry.java 代码如下。

【MyBeanImportRegistry.java】

```
package bean;
import org.springframework.beans.factory.support.BeanDefinitionRegistry;
import org.springframework.beans.factory.support.RootBeanDefinition;
import org.springframework.context.annotation.ImportBeanDefinitionRegistrar;
import org.springframework.core.type.AnnotationMetadata;
public class MyBeanImportRegistry implements ImportBeanDefinitionRegistrar {
    @Override
    public void registerBeanDefinitions(AnnotationMetadata importingClassMetadata,
    BeanDefinitionRegistry registry) {
        //创建 RootBeanDefinition 对象，并指定 bean 相关信息，如 bean 的类型、作用域等
        RootBeanDefinition rootBeanDefinition = new RootBeanDefinition(Bean2.class);
        //为 Bean2.class 类型的 Bean 注册 id 为 MyBean2
        registry.registerBeanDefinition("MyBean2",rootBeanDefinition);
    }
}
```

修改 SpringDemoConfig.java 配置文件如下：

```
/*@Import(Bean2.class)*/
/*@Import(MyBeanImportSelector.class)*/
@Import(MyBeanImportRegistry.class)
```

在类上注释 MyBeanImportSelector 类,导入 MyBeanImportRegistry 类。
修改测试类 ImportTest.java,将其中获取 Bean2 对象的语句改为以下代码:

```
Bean2 bean2=(Bean2)ctx.getBean("MyBean2")
```

运行 ImportTest.java,也能看到如图 2-13 所示的运行结果。
以上三种用法可以混合在一个 @Import 中使用。例如可以写成:

```
@Import({Bean2.class,MyBeanImportSelector.class,MyBeanImportSelector.class})
```

其中,前两种方法的 Bean 都是以全类名的方式注册,而第三种可自定义 Bean 的 id。

2.1.4 通过 FactoryBean 工厂创建 Bean

此方法创建一个实现了 FactoryBean 接口的 Java 类。在 Java 类中覆写两个方法 getObject 和 getObjectType,getObject 方法用于返回 Bean 对象。getObjectType 用于返回 Bean 对象的类型。下面还是以创建 Bean2 对象为例说明此方法的具体应用。

在 bean 文件夹下新建 MyFactoryBean.java 类,该类继承 FactoryBean 接口,并覆写 getObject()方法,该方法返回 Bean2 对象。覆写 getObjectType 方法,该方法返回 Bean2 对象的类型 Bean2.class。MyFactoryBean.java 代码如下。

【MyFactoryBean.java】

```java
package bean;
import org.springframework.beans.factory.FactoryBean;
public class MyFactoryBean implements FactoryBean<Bean2> {
    @Override
    public Bean2 getObject() throws Exception {
        System.out.println("执行 getObject()方法创建 Bean2 对象");
        return new Bean2();
    }
    @Override
    public Class<?> getObjectType() {
        return Bean2.class;
    }
}
```

在 SpringDemoConfig.java 配置文件中添加如下代码,进行 MyFactoryBean Bean 的配置,并注释掉 SpringDemoConfig 类上的@Import(MyBeanImportSelector.class)注解。

```java
@Bean
public MyFactoryBean myFactoryBean(){
    return new MyFactoryBean();
}
```

修改测试类 ImportTest.java 的代码,如下面的粗体代码所示。其中,ctx.getBean ("myFactoryBean")获取的并不是 MyFactoryBean 对象,而是 getObject 方法返回的 Bean2

对象。

【ImportTest.java】

```java
import bean.Bean2;
import bean.MyFactoryBean;
import config.SpringDemoConfig;
import org.springframework.context.annotation.AnnotationConfigApplicationContext;
public class ImportTest {
    public static void main(String[] args){
        AnnotationConfigApplicationContext ctx=
            new AnnotationConfigApplicationContext(SpringDemoConfig.class);
        /*通过myFactoryBean获取到的是myFactoryBean工厂调用getObject创建的Bean2对象*/
        Bean2 bean2=(Bean2)ctx.getBean("myFactoryBean");
        bean2.say();
    }
}
```

运行结果如图 2-5 所示。程序确实调用了 getObject 方法执行 Bean2 的构造方法并创建 Bean。

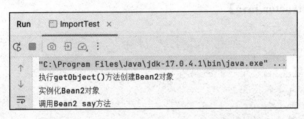

图 2-5　ImportTest.java 程序的运行结果 2

如果要获取 MyFactoryBean 工厂对象本身，则在 Bean 的 id 前加一个 & 即可。

```java
MyFactoryBean myFactoryBean=(MyFactoryBean)ctx.getBean("&myFactoryBean");
```

运行结果如图 2-6 所示，成功获取到 MyFactoryBean 工厂对象。

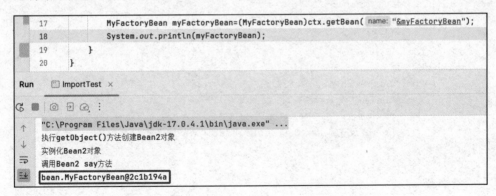

图 2-6　控制台打印 MyFactoryBean 工厂对象

任务 2.2　基于注解创建有变量属性的 Bean

在本项目的任务 2.1 中介绍了 4 种创建无变量属性 Bean 的方式，这 4 种方式同样可以应用于创建有变量属性的 Bean，只不过在创建时需要对成员变量进行赋值操作，这种赋值操作也称为依赖注入。依赖注入方式有很多种，可以通过调用有参构造方法注入，还可以利用 set 方法注入，甚至可以利用注解注入。注入值的类型可以多种多样，例如可以为简单数据类型、集合甚至是其他的 Bean 对象。下面介绍如何使用不同方式进行依赖注入。

2.2.1　利用有参构造方法注入

如果要利用有参构造方法注入，Bean 对应的类中必须提供有参构造方法，在有参构造方法中对参数进行赋值，具体用法如下。

在 bean 文件夹下创建 BeanWithConstructor.java 文件，在 BeanWithConstructor 内部定义 3 个属性，并添加有参构造方法。3 个属性分别对应不同的三种类型（简单数据类型、java 对象、引用数据类型）。同时添加 toString 方法用于打印内部属性值。

【BeanWithConstructor.java】

```java
package bean;
import java.util.List;
public class BeanWithConstructor {
    //参数类型为简单数据类型
    private String name;
    //参数类型为其他 Bean 对象
    private Bean1 bean1;
    //参数类型为引用数据类型，如 List
    private  List<String> stringList;
    //有参构造方法注入
    public BeanWithConstructor(String name, Bean1 bean1 , List<String> stringList){
        this.name=name;
        this.bean1=bean1;
        this.stringList=stringList;
    }
    @Override
    public String toString() {
        return "BeanWithParameter{" +
            "name=" + name +
            ", bean1='" + bean1 + '\'' +
            ", stringList='" + stringList + '\'' +
            '}';
    }
}
```

在 SpringDemoConfig.java 文件中注释掉 getBean2 和 getBean3 两个方法，使 Bean1 唯一，否则 Bean1 注入不成功。同时添加以下代码，配置生成 BeanWithConstructor 类型的

Bean。

```
@Bean
public BeanWithConstructor beanWithConstructor(Bean1 bean1){
    List<String> stringList=Stream.of("aa","bb","cc").toList();
    return new BeanWithConstructor("my",bean1,stringList);
}
```

新建测试类 BeanWithConstructorTest.java，获取 BeanWithConstructor 对象，看值是否注入成功。

【BeanWithConstructorTest.java】

```
import bean.BeanWithConstructor;
import config.SpringDemoConfig;
import org.springframework.context.annotation.AnnotationConfigApplicationContext;
public class BeanWithConstructorTest {
    public static void main(String [] args){
        AnnotationConfigApplicationContext ctx=
        new AnnotationConfigApplicationContext(SpringDemoConfig.class);
        BeanWithConstructor beanWithConstructor=
        (BeanWithConstructor)ctx.getBean("beanWithConstructor");
        System.out.println(beanWithConstructor);
    }
}
```

运行结果如图 2-7 所示，可以看到成功获取到 BeanWithConstructor 中的三个属性注入的值。

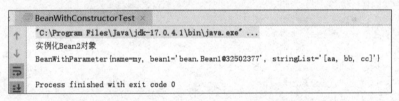

图 2-7　BeanWithConstructorTest.java 程序的运行结果

2.2.2　利用 set 方法注入

用 set 方法对参数进行设值注入，要求 Bean 类必须提供各个参数对应的 set 方法，具体用法如下。

在 bean 文件夹下创建 BeanWithSet.java 文件，在 BeanWithSet 内部定义 3 个属性，并添加相应的 get 和 set 方法。3 个属性分别对应不同的三种类型（简单数据类型、java 对象、引用数据类型）。同时添加 toString 方法，用于打印内部属性值。

【BeanWithSet.java】

```
package bean;
import java.util.List;
```

```java
public class BeanWithSet {
    //参数类型为简单数据类型
    private String name;
    //参数类型为其他 Bean 对象
    private Bean1 bean1;
    //参数类型为引用数据类型,如 List
    private List<String> stringList;
    //提供 name 属性的 set 方法
    public void setName(String name) {
        this.name = name;
    }
    //提供 bean1 属性的 set 方法
    public void setBean1(Bean1 bean1) {
        this.bean1 = bean1;
    }
    //提供 stringList 属性的 set 方法
    public void setStringList(List<String> stringList) {
        this.stringList = stringList;
    }
    public String getName() {
        return name;
    }
    public Bean1 getBean1() {
        return bean1;
    }
    public List<String> getStringList() {
        return stringList;
    }
    @Override
    public String toString() {
        return "BeanWithSet{" +
                "name=" + name +
                ", bean1='" + bean1 + '\'' +
                ", stringList='" + stringList + '\'' +
                '}';
    }
}
```

在 SpringDemoConfig.java 文件中添加以下代码,配置 BeanWithSet Bean。

```java
@Bean
public BeanWithSet beanWithSet(Bean1 bean1){
    List<String> stringList=Stream.of("aa","bb","cc").toList();
    BeanWithSet beanWithSet=new BeanWithSet();
    beanWithSet.setName("my");
    beanWithSet.setBean1(bean1);
    beanWithSet.setStringList(stringList);
    return beanWithSet;
}
```

新建测试类 BeanWithSetTest.java，获取 BeanWithSet 对象，看值是否注入成功。
【BeanWithSetTest.java】

```
import bean.BeanWithSet;
import config.SpringDemoConfig;
import org.springframework.context.annotation.AnnotationConfigApplicationContext;
public class BeanWithSetTest {
    public static void main(String [] args){
        AnnotationConfigApplicationContext ctx=
        new AnnotationConfigApplicationContext(SpringDemoConfig.class);
        BeanWithSet beanWithSet=(BeanWithSet)ctx.getBean("beanWithSet");
        System.out.println(beanWithSet);
    }
}
```

运行结果如图 2-8 所示，可以看到成功获取到 BeanWithSet 的三个属性注入的值。

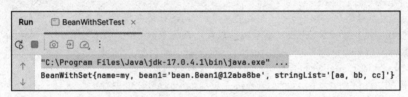

图 2-8　BeanWithSetTest.java 程序的运行结果

2.2.3　利用注解注入

Spring 可以利用注解方式进行值的注入，常用的有 @Value、@Resource、@Autowired 等，下面分别介绍。

1. @Value 注解

@Value 通常用于为当前 Bean 的属性注入外部数据值，并自动进行类型转换。@Value 可注入系统参数数据、环境变量数据和外部配置文件数据。除注入简单数据类型外，@Value 还能注入引用数据类型数据甚至其他 Bean。当 @Value 注解修饰属性时，该属性可以省略 set 方法。@Value 注解有如下三种使用方式。

(1) @Value("常量")。该方式表示为属性注入一个常量值。如 @Value("zhangsan") 表示为属性注入一个常量值 zhangsan。

(2) @Value(${key:defaultvalue})。该方式表示使用外部配置文件中键为 key 对应的值注入属性，如果找不到，则注入默认值 defaultvalue，默认值 defaultvalue 可省略，一般写成 @Value(${property})。如 @Value(${url:127.0.0.1})，读取外部配置文件中键为 url 对应的值注入属性，如果找不到，使用 127.0.0.1 默认值注入。

(3) @Value(#{ele?:defaultvalue})。该方式使用 EL 表达式为属性注入值，如果找不到则注入默认值 defaultvalue，默认值 defaultvalue 可省略，一般写成 @Value(#{ele})。其中，ele 可以是常量，也可以是一个自定义对象，还可以是自定义对象或环境变量中的某个属性。例如，

@Value("#{systemEnvironment['JAVA_HOME']}:null")

使用环境变量 JAVA_HOME 的路径注入属性，如果找不到，使用 null 值注入。

下面说明@Value 的具体用法。

在 Bean 目录下新建 BeanWithValue.java，在其中定义多个属性，用于注入不同类型的外部数据，并添加 toString 方法。其中，name 属性注入常量值，systemProperties 属性注入系统参数 Java 的版本号，systemEnvironment 属性注入环境变量 JAVA_HOME，bean1 属性注入其他 Bean 对象，stringList 属性注入其他 Bean 的属性值。

【BeanWithValue.java】

```java
package bean;import org.springframework.beans.factory.annotation.Value;
import java.util.List;
public class BeanWithValue {
    //参数类型为基本数据类型,注入常量值
    @Value("my")
    private String name;
    /* 参数类型为基本数据类型,注入值从系统参数、环境变量、外部配置文件获取 */
    //使用 systemProperties 从系统参数注入系统参数 Java 版本号
    @Value("#{systemProperties['java.version']}")
    private String systemProperties;
    //使用 systemEnvironment 从环境变量注入 JAVA_HOME 路径
    @Value("#{systemEnvironment['JAVA_HOME']?:0}")
    private String systemEnvironment;
    //注入其他 Bean 对象,beanWithSet 为 Bean 的 id
    @Value("#{beanWithSet}")
    private BeanWithSet beanWithSet;
    //参数类型为自定义的引用数据类型,如 List<String>,注入值从其他 Bean 对象属性获取
    //beanWithSet 为 Bean 的 id,stringList 为该 Bean 属性
    @Value("#{beanWithSet.stringList}")
    private List<String> stringList;
    @Override
    public String toString() {
        return "BeanWithValue{" +'\n'+
            "name=" + name +","+'\n'+
            "systemProperties='" +","+ systemProperties + '\n'+
            "systemEnvironment='" +","+ systemEnvironment +'\n'+
            "beanWithSet='" + beanWithSet +","+'\n'+
            "stringList='" + stringList +
            '}';
    }
}
```

在 SpringDemoConfig.java 文件最后添加以下 BeanWithValue Bean 的配置。

```java
@Bean
public BeanWithValue beanWithValue(){
    BeanWithValue beanWithValue=new BeanWithValue();
```

```
        return beanWithValue;
    }
```

新建测试类 BeanWithValueTest.java,获取 BeanWithValue 对象,看值是否注入成功。

【BeanWithValueTest.java】

```
import bean.BeanWithValue;
import config.SpringDemoConfig;
import org.springframework.context.annotation.AnnotationConfigApplicationContext;
public class BeanWithValueTest {
    public static void main(String [] args){
        AnnotationConfigApplicationContext ctx=
            new AnnotationConfigApplicationContext(SpringDemoConfig.class);
        BeanWithValue beanWithValue=(BeanWithValue)ctx.getBean("beanWithValue");
        System.out.println(beanWithValue);
    }
}
```

运行结果如图 2-9 所示,可以看到 BeanWithValue 的所有属性成功获取到注入的值。

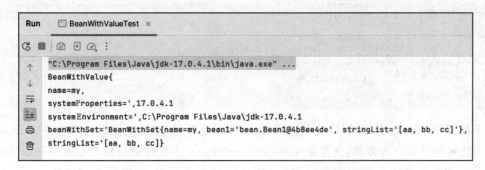

图 2-9　BeanWithValueTest 程序的运行结果

总之,@Value 使用非常灵活,一般情况下如果要注入配置文件中数据,使用 ${};如果要注入一个对象或者对象的属性,使用 #{};如果注入自定义复杂数据类型,使用 #{}和 ${}组合。

2. @Resource 注解和@Autowired 注解

在 Web 开发的 MVC 三层结构中,需要不同层级的 Bean 互相注入。例如把持久层 Bean 注入业务层属性,把业务层 Bean 注入控制层属性中。一般利用@Resource、@Autowired 注解进行注入。@Resource、@Autowired 注解可以写在属性上,也可以写在 set 方法上。如果写在属性上,则 set 方法可以省略。

下面介绍@Resource、@Autowired 注解的详细使用。

(1) @Resource 注解。@Resource 并不是 Spring 的注解,但是 Spring 支持该注解的注入。@Resource 注解有两个重要属性为 name 和 type,其中,name 属性为 Bean 的名字,type 属性为 Bean 的类型。使用@Resource 注入时,如果不指定 name 和 type,则默认通过属性名匹配 Bean 注入,如果找不到,再通过类型匹配。如果仅指定了 name,则通过属性名匹配 Bean 注入,找不到则报错。如果仅指定了 type,则通过属性类型匹配 Bean 自动注入。

如果同时指定了 name 和 type,则通过属性名和类型精确匹配 Bean。

Spring 5 中使用@Resource 注解注入时需要导入 javax.annotation 包。而在 Spring 6 中,javax 命名空间被 jakarta 代替,因此,在 Spring 6 中使用@Resource 注解需要导入 jakarta.annotation 包。

在 Maven 中央仓库中搜索 jakarta.annotation,搜索结果如图 2-10 所示,当前最新版本是 2.1.1。

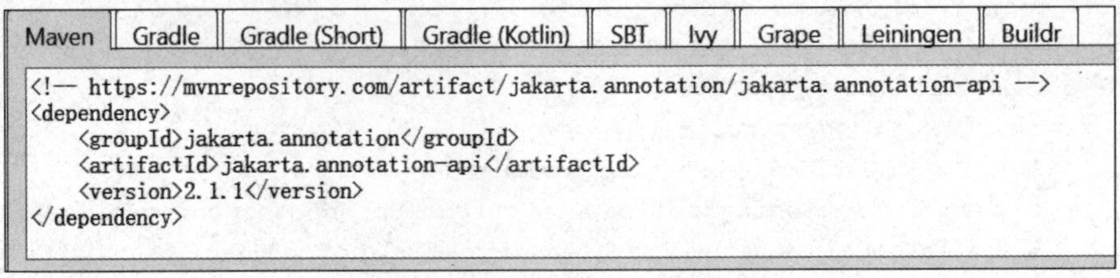

```
<!-- https://mvnrepository.com/artifact/jakarta.annotation/jakarta.annotation-api -->
<dependency>
    <groupId>jakarta.annotation</groupId>
    <artifactId>jakarta.annotation-api</artifactId>
    <version>2.1.1</version>
</dependency>
```

图 2-10　jakarta.annotation 依赖

需先在 SpringDemo2 的 pom.xml 文件中添加如下代码,引入 jakarta.annotation 依赖,然后才能在程序中使用@Resource 注解。

```xml
<!-- 配置 jakarta.annotation 依赖-->
<dependency>
    <groupId>jakarta.annotation</groupId>
    <artifactId>jakarta.annotation-api</artifactId>
    <version>2.1.1</version>
</dependency>
```

下面利用@Resource 注解将持久层的 Bean 注入业务层,在测试类中调用业务层 Bean 对象的方法打印结果。为方便,这里采用@ComponentScan 注解自动扫描并结合@Repository、@Service 注解创建持久层和业务层 Bean,持久层和业务层 Bean 之间通过@Resource 注解注入。

在 dao 层目录下新建一个接口类 DataDao.java,内部定义一个 getData 方法,用于模拟获取数据库数据。该类有 2 个实现类 DataDaoImpl1.java 和 DataDaoImpl2.java。DataDaoImpl1.java 和 DataDaoImpl2.java 都使用@Repository 注解标记为一个持久层的 Bean。

【DataDao.java】

```java
package dao;
public interface DataDao {
    public void getData();
}
```

【DataDaoImpl1.java】

```java
package dao;
import org.springframework.stereotype.Repository;
@Repository
```

```
public class DataDaoImpl1 implements DataDao{
    @Override
    public void getData() {
        System.out.println("调用 DataDaoImpl1 的 getData 方法获取数据");
    }
}
```

【DataDaoImpl2.java】

```
package dao;
import org.springframework.stereotype.Repository;
@Repository
public class DataDaoImpl2 implements DataDao{
    @Override
    public void getData() {
        System.out.println("调用 DataDaoImpl2 的 getData 方法获取数据");
    }
}
```

在 src/main/java 目录下新建 service 目录。在 service 目录新建接口类 DataService.java 和实现类 DataServiceImpl.java，在其中定义私有变量 dataDao 并使用@Resource 注解注入 DataDao 类型的 Bean。DataServiceImpl 内部调用 dataDao 对象的 getData 方法。这里直接在属性上添加@Resource 注解，不指定 name 和 type。

【DataService.java】

```
package service;
import dao.DataSource;
public interface DataService {
    public void getData() ;
}
```

【DataServiceImpl.java】

```
package service;
import dao.DataDao;
import dao.DataSource;
import jakarta.annotation.Resource;
import org.springframework.stereotype.Service;
@Service
public class DataServiceImpl implements DataService {
    @Resource
    private DataDao dataDao;
    @Override
    public void getData() {
        dataDao.getData();
    }
}
```

在 SpringDemoConfig 配置类上添加以下注解：

```
@ComponentScan({"dao","service"})
```

用于自动扫描 dao 文件夹和 service 文件夹下标识的 Bean。

新建测试类 DataServiceTest.java，内部访问 DataService Bean 并调用 getData 方法获取数据。

【DataServiceTest.java】

```
import config.SpringDemoConfig;
import org.springframework.context.annotation.AnnotationConfigApplicationContext;
import service.DataService;
public class DataServiceTest {
    public static void main(String [] args){
        AnnotationConfigApplicationContext ctx=
        new AnnotationConfigApplicationContext(SpringDemoConfig.class);
        DataService dataService=(DataService)ctx.getBean("dataServiceImpl");
        dataService.getData();
    }
}
```

运行 DataServiceTest.java，出现如下所示的报错信息。

```
org.springframework.beans.factory.NoUniqueBeanDefinitionException:
No qualifying bean of type 'dao.DataDao' available: expected single matching
bean but found 2: dataDaoImpl1,dataDaoImpl2
```

报错信息提示用户发现了 2 个 DataDao 类型的 Bean，分别是 dataDaoImpl1 和 dataDaoImpl2。由于 DataDao 类型的 Bean 不唯一，因此无法注入。这是因为@Resource 未设置 name 和 type，默认按照属性名 dataDao 注入，但是容器中并没有名为 dataDao 的 Bean，就通过类型 DataDao 去查找 Bean，结果发现有 2 个 DataDao 类型的 Bean 名字分别是 DataDaoImpl1 和 DataDaoImpl2，Bean 不唯一就报错。

要解决这个问题，只需在使用@Resource 时指定 name 或者同时指定 name 和 type。以下两种方式都可以。

```
@Resource(name="dataDaoImpl1")
@Resource(name="dataDaoImpl1",type=DataDao.class)
```

再次运行 DataServiceTest.java，则程序正常运行，运行结果如图 2-11 所示。

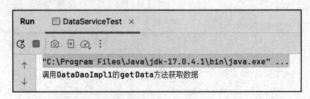

图 2-11　DataServiceTest 程序的运行结果

（2）@Autowired 注解。@Autowired 注解默认通过属性类型匹配 Bean 注入。使用方法和@Resource 类似，直接在属性上添加修饰即可。但是@Autowired 内部没有 name 属性设置，使用时如果有多个类型一样的 Bean 注入，则需要和@Qualifier 注解组合使用来限定唯一的 Bean，否则会报错。如果直接将 DataServiceImpl.java 里面的@Resource 注解改成@Autowired，再次运行 DataServiceTest.java。程序就会出现报错信息，因为通过类型匹配存在 2 个同类型的 Bean，Bean 不唯一。

此时可以将@Autowired 和@Qualifier 注解组合使用，将 DataServiceImpl 私有变量 dataDao 上面的注解修改如下。

```
//@Qualifier 和@Autowired 组合使用，@Qualifier 限定匹配的 Bean id
@Autowired
@Qualifier(value="dataDaoImpl1")
private DataDao dataDao;
```

再次运行 DataServiceTest.java，程序执行得到正常的运行结果。

此外，@Autowired 内部有一个属性 required，用于指定注入对象是否存在，required 默认值为 true，要求注入的 Bean 对象必须存在。如果注入对象允许为 null 值，则可以设置 required 为 false。

任务 2.3　了解 Bean 的作用域

在本项目的任务 2.1 中提到@Bean 注解可以结合@Scope 注解在实例化 Bean 时指定 Bean 的作用域范围。下面将对 Bean 的作用域做详细介绍。

2.3.1　初识 Bean 作用域

Spring 提供了 7 个 bean 的作用域，其中 4 个仅在 WebApplicationContext 中可用。详细信息如表 2-3 所示。

表 2-3　Spring Bean 的作用域

作用域	描　　述
singleton	使用 singleton 定义 Bean 时，Spring 容器仅创建一个 Bean 实例，每次返回的是同一个 Bean，即单例模式
prototype	使用 prototype 定义 Bean 时，Spring 容器可以创建多个 Bean 实例，每次返回的都是一个新的 Bean，即多例模式
request	request 仅对 HTTP 请求产生作用，使用 request 定义 Bean 时，每次 HTTP 请求都会创建一个新的 Bean，仅在 WebApplicationContext 中可用
session	session 仅用于 HTTP 会话中，同一个会话共享一个 Bean。不同会话使用不同的 Bean，仅在 WebApplicationContext 中可用

续表

作用域	描述
global-session	global-session 仅用于 HTTP Session，同 session 作用域不同的是，所有的会话共享一个 Bean，仅在 WebApplicationContext 中可用
application	为每个 ServletContext 对象创建一个 Bean，仅在 WebApplicationContext 中可用
websocket	为每个 WebSocket 对象创建一个 Bean，仅在 WebApplicationContext 中可用

其中，request、session、global-session、application、websocket 这些作用域在 Web 开发中使用，这里不详细讲解。Spring 默认生成的 Bean 的作用域都是单例，如果要使生成的 Bean 作用域为多例，只需在生成 Bean 的方法上添加注解@Scope("prototype")即可。

2.3.2 Bean 的作用域与线程安全

Java 类可以分为两类，分别是无状态类和有状态类。无状态类指的是类的内部状态在整个生命期间都不会发生变化，即类内部没有属性存储数据或者内部属性值为另一个无状态对象。无状态类有如下两种形式。

(1) 类内部没有定义属性。如下面定义的 NotStateful 类。

```
public class NotStateful{
}
```

(2) 类内部定义了属性，但属性值也是一个无状态对象。如 UserServiceImpl 类中一般会注入 Userdao 实例对象，但是 Userdao 类内部一般没有定义属性存储具体数据，只有一些数据操作方法。因此，Userdao 是无状态类，Userdao 类的实例对象就是无状态对象，UserServiceImpl 自然就是无状态类。

```
public class UserServiceImpl{
    private Userdao userdao;
}
```

无状态类创建的 Bean 对象被称为无状态对象，可直接通过调用类默认无参构造方法创建 Bean 对象。多线程访问无状态对象时，所有线程访问同一个对象，由于对象内部没有数据存储，不会造成数据污染。因此，多线程访问无状态对象，无论对象是单例模式还是多例模式，都是线程安全的。

有状态类指的是 Java 类内部通过定义变量来存储具体的数据。以下的 StatefulBean 类就是一个有状态类。

```
public class StatefulBean {
    private String name;
    public String getName() {
        return name;
    }
    public void setName(String name) {
```

```
        this.name = name;
    }
}
```

有状态类创建的 Bean 对象被称为有状态对象,可通过调用类有参构造方法创建 Bean 对象。多线程访问有状态对象时,如果 Bean 设置为单例,所有线程访问同一个对象,由于对象内部需存储各线程的数据,会造成资源竞争,导致数据污染。如果 Bean 设置为多例,线程各自使用一份 Bean 对象存储各自数据,则是线程安全的。

实际上在 Spring 中,Bean 无论是否有默认状态,都是单例的。

2.3.3 Spring 中单例 Bean 的多线程访问控制

在 Java 开发中,有两种方式解决多线程下单例 Bean 的线程安全问题,分别是线程同步机制和 ThreadLocal。

线程同步机制是通过对象的锁机制保证同一时间只有一个线程访问对象中的变量,其他线程排队等待。使用线程同步机制要求开发人员自主控制对象中的变量进行读写、锁定对象、释放对象锁等操作,程序设计和编写难度相对较大。

ThreadLocal 为每个线程提供一个独立的变量副本,来隔离多个线程对数据的访问冲突。开发人员可以把对象内不安全的变量都封装进 ThreadLocal 中,程序设计和编写难度相对较低。因此,在多线程环境下,ThreadLocal 使用时比线程同步机制更加便捷。

Spring 内部就是使用 ThreadLocal 来解决线程安全问题,这样即使 Bean 是有状态的单例 Bean,Spring 也能通过使用 ThreadLocal 进行处理,保证有状态单例 Bean 访问的线程安全。

任务 2.4　了解 Bean 的生命周期

在 Spring 中,Bean 的生命周期是一个很复杂的执行过程,始于 Bean 的实例化,止于 Bean 的销毁。学习 Bean 的生命周期的意义在于开发人员可以在 Bean 实例化后及被销毁前这段时间执行一些自己所需的相关操作。

Spring 容器只管理单例 Bean 的生命周期。对于多例 Bean 而言,Spring 只负责创建 Bean 实例,当 Bean 实例创建完成后,就交给客户端代码管理,Spring 容器将不再管理其生命周期。

在 Spring 容器中,一个 Bean 从能够被正常使用到被销毁需要进行很多工作。这些工作共同组成了 Bean 的生命周期。Bean 的生命周期如下所示。

Bean 的生命周期

任务2.5 综合案例：统计用户登录次数

在现代的Java开发中，Spring的Bean被大量运用于编程当中。基于前面介绍的Spring Bean相关知识，这里以统计用户登录次数为例，演示Spring注解创建Bean的综合应用。

2.5.1 案例任务

该任务模拟服务器对登录的用户次数进行统计。要求新建User类，类内部定义username和login_count属性，所有登录用户信息利用一个集合保存。用户登录需根据用户名判断是否为新用户。如果为新用户，则设置登录次数初始为1，将该用户添加到集合中，否则对集合中相应用户的登录次数加1。整个功能可利用MVC三层结构实现Bean的依赖注入。

2.5.2 任务分析

按照任务要求分析，该任务实现思路如下。

（1）整个代码结构分为Dao、Service、Controller三层结构，其中Dao和Service层需新建接口和实现类，并定义相关方法。Controller层新建类并定义相应方法，用于接收外部用户登录信息。

（2）在Dao层实现类中需定义集合，用于存储用户数据；定义数据操作方法，用于向集合新增数据和更新数据；定义方法，获取所有用户登录信息并查看保存数据是否正确。

（3）在Service层注入Dao层对象，Service层需根据用户是否为新用户调用Dao层不同方法实现业务逻辑。

（4）在Controller层注入Service层对象，Controller层定义相应方法接收外部用户登录信息并调用Service层方法。

2.5.3 任务实施

在SpringDemo2的bean文件夹下新建User.java类，类中定义user name和login_name属性并提供对应的get和set方法。在SpringDemo2的dao文件夹下新建LoginDao.java接口和LoginDaoImpl.java实现类。在LoginDaoImpl类中定义集合userlist以存储所有登录用户信息，定义adduserLoginCount方法来添加新用户登录信息，定义updateUserLoginCount方法更新已存在用户的登录信息，定义findUserByUserName方法判断用户是否为新用户，定义findAllUserInfo方法返回userlist存储的所有用户信息。LoginDaoImpl类利用@Repository注解标识为持久层Bean。

【LoginDao.java】

```
package dao;
import bean.User;
```

```java
import java.util.ArrayList;
public interface LoginDao {
    void addUserLoginCount(String username);
    void updateUserLoginCount(String username);
    boolean findUserByUserName(String username);
    ArrayList<User> findAllUserInfo();
}
```

【LoginDaoImpl.java】

```java
package dao;
import bean.User;
import org.springframework.stereotype.Repository;
import java.util.ArrayList;
import java.util.stream.Collectors;
@Repository
public class LoginDaoImpl implements LoginDao {
    //userlist存储所有用户登录信息
    private static ArrayList<User> userlist;
    static{
        userlist=new ArrayList<User>();
    }
    //添加新用户的登录信息,登录次数为1
    @Override
    public void addUserLoginCount(String username) {
        User user=new User();
        user.setUsername(username);
        user.setLogin_count(1);
        userlist.add(user);
    }
    //对于已登录用户,要再次登录,登录次数加1
    @Override
    public void updateUserLoginCount(String username) {
        userlist.stream().map(s->{
            if(s.getUsername().equals(username)){
                s.setLogin_count(s.getLogin_count()+1);
            }
            return s;
        }).collect(Collectors.toList());
    }
    //判断用户是否为新用户
    @Override
    public boolean findUserByUserName(String username) {
        //anyMatch用于在流中匹配数据
         boolean flag= userlist.stream().anyMatch(s->s.getUsername().equals(username));
        return flag;
    }
    //获取userlist存储所有用户登录信息
```

```
    @Override
    public ArrayList<User> findAllUserInfo() {
        return userlist;
    }
}
```

在 SpringDemo2 的 service 文件夹下新建 LoginService.java 接口和 LoginServiceImpl.java 实现类。在 LoginServiceImpl 类中利用@Autowired 注解注入 LoginDao 对象。类中定义 updateUserLoginInfo 方法根据登录用户是否为新用户，调用相应方法进行处理。定义 findAllUserInfo 方法，调用类中 findAllUserInfo 方法，返回 userlist 存储的所有用户登录信息。LoginServiceImpl 利用@Service 注解标识为业务层 Bean。

【LoginService.java】

```
package service;
import bean.User;
import java.util.ArrayList;
public interface LoginService {
    void updateUserLoginInfo(String username);
    ArrayList<User> findAllUserInfo();
}
```

【LoginServiceImpl】

```
package service;
import bean.User;
import dao.LoginDao;
import org.springframework.beans.factory.annotation.Autowired;
import org.springframework.stereotype.Service;
import java.util.ArrayList;
@Service
public class LoginServiceImpl implements LoginService {
    @Autowired
    private LoginDao loginDao;
    //判断登录用户是否为新用户，如是新用户则执行 addUserLoginCount,否则执行
    //updateUserLoginCount
    @Override
    public void updateUserLoginInfo(String username){
        if(loginDao.findUserByUserName(username)){
            loginDao.updateUserLoginCount(username);
        }else{
            loginDao.addUserLoginCount(username);
        }
    }
    //获取 userlist 存储的所有用户登录信息
    @Override
    public ArrayList<User> findAllUserInfo() {
        return loginDao.findAllUserInfo();
    }
}
```

在 src/main/java 目录下新建 controller 文件夹,在该文件夹中新建 LoginController.java,在 LoginController 类中利用@Autowired 注入 LoginService 对象,定义方法接收登录的所有用户信息。为方便登录的用户信息使用 List 集合传入,定义 findAllUserInfo 方法,内部调用 LoginService 对象的 findAllUserInfo 方法,返回 userlist 存储的所有用户登录信息,以检验用户登录数据是否正确存储。LoginController 利用@Controller 注解标识为控制层 Bean。

【LoginController.java】

```java
package controller;
import bean.User;
import org.springframework.beans.factory.annotation.Autowired;
import org.springframework.stereotype.Controller;
import service.LoginService;
import java.util.ArrayList;
import java.util.List;
@Controller
public class LoginController {
    @Autowired
    private LoginService loginService;
    public void addUser(List<String> userlist){
        userlist.stream().forEach(x->{
            loginService.updateUserLoginInfo(x);
        }
        );
    }
    public ArrayList<User> findAllUserInfo(){
        return loginService.findAllUserInfo();
    }
}
```

修改配置类型 SpringDemoConfig.java 文件,将自动扫描路径设为如下:

```java
@ComponentScan({"dao","service","controller"})
```

编写测试类 LoginControllerTest.java,在其中获取 LoginController 对象并传入用户信息 userList,再调用 findAllUserInfo 方法检查数据是否正确存储。这里假设输入登录用户集合为{"aaa","bbb","eee","ddd","bbb"}。如果程序正常,userList 中存储的数据应为 aaa、ddd、eee 用户登录 1 次,bbb 用户登录 2 次。

【LoginControllerTest.java】

```java
import bean.User;
import config.SpringDemoConfig;
import controller.LoginController;
import org.springframework.context.annotation.AnnotationConfigApplicationContext;
import java.util.ArrayList;
import java.util.List;
import java.util.stream.Stream;
```

```java
public class LoginControllerTest {
    public static void main(String [] args){
        AnnotationConfigApplicationContext ctx=
        new AnnotationConfigApplicationContext(SpringDemoConfig.class);
        LoginController loginController=(LoginController)ctx.getBean("loginController");
        List<String> userList=Stream.of("aaa","bbb","bbb","ddd","eee").toList();
        loginController.addUser(userList);
        ArrayList<User> allUserInfo=loginController.findAllUserInfo();
        allUserInfo.forEach(x->System.out.println(x));
    }
}
```

运行结果如图 2-12 所示，可以看到控制台打印的结果是正确的。

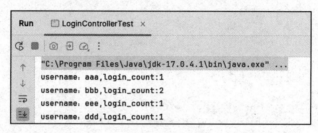

图 2-12 LoginControllerTest 程序的运行结果

小　　结

本章详细介绍了 Spring 中 Bean 的创建和属性值的注入、Bean 的作用域与线程安全、Bean 的生命周期等内容。最后以一个综合案例——对用户登录次数进行统计，演示了在 MVC 三层架构下 Bean 的配置和使用。使读者对 Spring 中 Bean 的相关知识有更深入的认识。

课后练习：校验并分类统计登录用户信息

在任务 2.5 基础上做进一步改进，登录用户名以英文开头且长度为 3 的为合法用户，除此之外为非法用户。用集合分别对合法用户和非法用户登录数据进行分类存储，编程实现用控制台打印每个合法用户的登录次数和每个非法用户的登录次数。

项目 3　Spring AOP 编 程

AOP 全称为 aspect-oriented programming，即面向切面编程。面向切面编程是一种横向切入的编程思想，是对面向对象编程思想的一种补充，被大量运用于代码编写中。Spring 中也实现了相应的 AOP 框架，该框架的底层是通过 Java 的代理机制实现的。因此，本项目首先介绍代理相关知识，进而介绍 Spring AOP 的具体实现方式和编程应用。

任务 3.1　了解代理机制

代理是一种软件设计模式，目的在于实现代码复用。代理是用一个类间接执行另一个类中的方法或引用其成员。代理的例子在生活中随处可见。例如，房东要租房，一般会找中介帮忙挂出房源，如果有人租房，中介可以帮忙出租，中介就是一个代理。使用代理的好处在于租房过程中很多事情中介可以代劳，房东最终只关注房子以什么价位租出去，自己不需要去做太多的事情，如自己询问有没有租房者等。同时，租房者也不需要自己去联系房东租房，只需要找中介就能租房。总之，代理就是代理你办事的人或其他机构。

代理机制

下面以租房为例，介绍代理模式中涉及的一些基本要素，如图 3-1 所示。

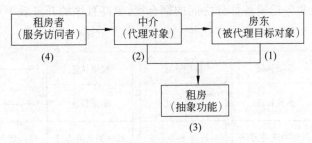

图 3-1　代理模式中基本元素

(1) 被代理目标对象：房东需要租房，但一般委托中介代理租房，房东就成为被代理对象，即目标对象。

(2) 代理对象：中介代理房东租房，中介就成为代理对象，与房东形成代理关系。

(3) 抽象功能：这里之所以说是抽象功能，是因为中介和房东实现的功能都是租房，只不过房东租房仅仅要做的就是租房，而中介除了帮房东实现租房外，还有一些附加事项要做，例如看房、收中介费、签合同、协调价格等。

(4) 服务访问者：找中介租房的人。

在 Java 开发中，代理分为静态代理和动态代理两种模式。静态代理模式是一个代理对象对应一个目标对象，代理关系是固定的，代理对象和目标对象一一绑定，动态代理模式下代理对象可以同时对应多个目标对象，代理对象和目标对象不是一一绑定，代理关系不是固定的。静态代理就好比中介代理一个房东的业务，在代理之前中介就已经知道要代理哪个房东的业务了。动态代理就好比中介可以同时服务多个房东的业务，只有租客来租房了，看中了哪个房东的房子，中介就代理该房东的业务。显然动态代理相比静态代理更加灵活，更有优势。

任务 3.2　初识 Spring AOP

动态代理就是 AOP 的底层实现。AOP 的宗旨就是为了保证开发者在不修改源代码的前提下，为系统中的业务组件添加某些非业务通用功能。使开发人员可以集中处理纯粹的业务，减少非业务代码对业务代码的侵入，增强代码的可读性和可维护性。

3.2.1　AOP 简介

在业务处理代码中，通常都会进行权限控制、日志记录、事务处理等操作。以传统方式来实现需要在每个业务代码中都添加相应的方法，这样势必会引起原有业务代码的改动，如果后续还需增加其他操作，又要再次修改所有业务代码。这不但增加了开发人员的工作量，同时使业务代码不再纯粹于业务，附加了一些非功能性的其他操作，增强了代码的耦合度。

为了解决这一问题，面向切面编程（AOP）思想随之产生。AOP 采取横向抽取机制，如图 3-2 所示。将分散在各个方法中的重复代码提取出来，然后在程序编译或运行时，再将这些提取出来的代码切入需要执行的地方。这种横向抽取机制底层正是通过动态代理实现的。AOP 能够降低业务逻辑各部分之间的耦合度，提高程序的开发效率和可重用性。

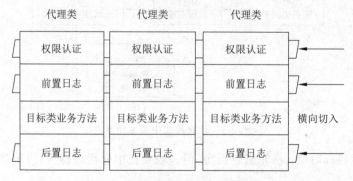

图 3-2　面向切面编程横向抽取机制

基于 Java 的代理机制,Spring 采用两种方式实现 AOP,分别是 JDK 动态代理和 CGLib 动态代理。其中,JDK 动态代理是 Spring AOP 的默认选择。当目标类实现了接口时,Spring 默认使用 JDK 动态代理实现 AOP,也可配置强制使用 CGLIB 动态代理。当目标类没有实现接口时,则必须使用 CGLIB 动态代理实现 AOP。

3.2.2 AOP 术语

AOP 利用横向切入技术把应用软件分为两部分:核心关注点和横切关注点。核心关注点就是与业务相关的处理流程。横切关注点就是跨越多个类且与业务无关的通用功能或方法,如权限、日志、事务等。横切关注点经常会发生在多个核心关注点中,如各业务流程方法可能都需要实现权限认证、写日志、事务管理等。在 AOP 中有许多术语,是学习 AOP 的基础,如图 3-3 所示。下面做具体介绍。

AOP 术语解释

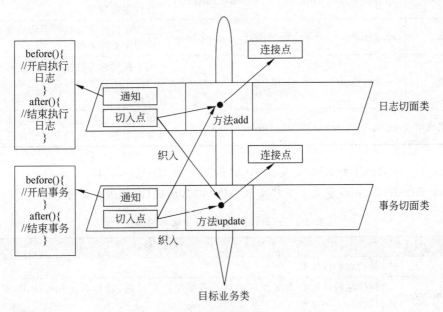

图 3-3 AOP 术语及实现过程

(1) 核心关注点:目标类中业务逻辑处理流程。

(2) 横切关注点:跨越多个类且与业务无关的通用功能或方法,如权限、日志、事务等。

(3) 切面类(aspect):模块化封装横切关注点通用功能形成的类。如权限模块切面类、日志模块切面类等。切面类中一般包含了切点和通知,指定了在何时何处完成通用功能的切入。

(4) 通知(advice):切面类中定义的方法,用于将横切关注点某些通用功能增强到目标类的方法中。

  (5) 目标(target)：被一个或多个切面通知的对象。

  (6) 代理(proxy)：向目标对象应用通知生成的对象。

  (7) 切入点(pointcut)：切面类中通知切入的地方。

  (8) 连接点(joinpoint)：应用程序执行时，切入点插入的位置。

  (9) 织入(weaving)：把切面应用到目标对象并创建代理对象的过程。

3.2.3 Spring 的两种 AOP 实现

  经过前面的学习，可以了解到 AOP 底层是基于动态代理机制实现的，动态代理对象是在程序运行时自动在内存中创建的。如果引入 Spring，这些动态代理对象的创建则交给 Spring 来做。Spring 利用 ProxyFactoryBean 类统一创建所有的动态代理对象并管理起来。Spring 提供了一些常用的 Advice API 用于增强目标方法。Spring 中常用的 Advice API 如表 3-1 所示。

表 3-1　Spring 中常用的 Advice API

API 类型	包 名	API 描 述
前置通知	org.springframework.aop.MethodBeforeAdvice	用于在目标方法执行前实施增强，如进行权限管理等
后置通知	org.springframework.aop.AfterReturningAdvice	用于在目标方法执行后实施增强，可以应用于关闭流等
环绕通知	org.aopalliance.intercept.MethodInterceptor	用于在目标方法执行前后实施增强，可以应用于方法执行前后记录日志、事务管理等
异常抛出通知	org.springframework.aop.ThrowsAdvice	用于在方法抛出异常后实施增强，可以应用于方法执行异常时记录日志

  ProxyFactoryBean 是 Spring 实现 AOP 的核心类，是 FactoryBean 接口的一个具体实现。ProxyFactoryBean 中定义了一些实现 AOP 的常用属性，如表 3-2 所示。

表 3-2　ProxyFactoryBean 中常用的属性

属 性	属性描述
target	被代理的目标类
proxyInterfaces	被代理的目标类要实现的接口。若不配置此属性，目标类使用 CGLIB 实现动态代理
proxyTargetClass	默认为 false。如果要代理目标类没有实现接口，则该属性为 true。目标类使用 CGLIB 实现动态代理
interceptorNames	需要插入目标类方法的通知
singleton	生成的代理对象是否为单例，默认为 true(单例)

  在 2.1.4 小节中我们已经学习过 FactoryBean 相关知识，了解到 FactoryBean 是可用于实例化 Bean，而 ProxyFactoryBean 用于为实例化的 Bean 创建代理对象。

  利用 ProxyFactoryBean 实现 AOP 有两种方法，下面将做具体介绍。

1. 基于 Spring Advice API 接口

此方法需先定义一个插入目标对象的类,该类根据实际需求继承表 3-1 中的某个 Spring Advice API 接口,并覆写相应的方法。然后在 Spring 配置类中利用 ProxyFactoryBean 配置相关属性实现 AOP。此方法的具体实现如下。

新建 SpringDemo3 项目,在 pom.xml 文件下添加 spring-webmvc 依赖,导入 Spring 相关依赖,其中也包含 AOP 依赖。同时配置<build>标签,设置 maven 编译插件,设置 Java 版本为 17。

【pom.xml】

```xml
<?xml version="1.0" encoding="UTF-8"?>
<project xmlns="http://maven.apache.org/POM/4.0.0"
    xmlns:xsi="http://www.w3.org/2001/XMLSchema-instance"
    xsi:schemaLocation="http://maven.apache.org/POM/4.0.0
    http://maven.apache.org/xsd/maven-4.0.0.xsd">
    <modelVersion>4.0.0</modelVersion>
    <groupId>SpringDemo3</groupId>
    <artifactId>SpringDemo3</artifactId>
    <version>1.0-SNAPSHOT</version>
    <dependencies>
        <dependency>
            <groupId>org.springframework</groupId>
            <artifactId>spring-webmvc</artifactId>
            <version>6.0.3</version>
        </dependency>
    </dependencies>
    <build>
        <plugins>
            <!-- 配置当前项目的 JDK 版本信息 -->
            <plugin>
                <groupId>org.apache.maven.plugins</groupId>
                <artifactId>maven-compiler-plugin</artifactId>
                <version>3.8.1</version>
                <configuration>
                    <!--LocalVariableTableParameterNameDiscoverer
                    在 spring6.x 中移除内容,需要为编译插件配置-parameters 参数,
否则执行 AOP 时会有警告-->
                    <compilerArgs>
                        <arg>-parameters</arg>
                    </compilerArgs>
                    <source>17</source>
                    <target>17</target>
                    <encoding>UTF-8</encoding>
                </configuration>
            </plugin>
        </plugins>
    </build>
</project>
```

在 SpringDemo3 的 src/main/java 目录下新建 aop 文件夹，在 aop 文件夹下新建业务层目标接口类 BirdInterface.java，内部提供 fly 方法。新建 Bird.java 类实现 BirdInterface 接口，并覆写 fly 方法。

【Bird.java】

```
package aop;
import org.springframework.stereotype.Service;
@Service
public class Bird implements BirdInterface{
    public void fly(){
        System.out.println("鸟在飞行");
    }
}
```

【BirdInterface.java】

```
package aop;
public interface BirdInterface {
    void fly();
}
```

在 aop 文件夹下新建前置通知类 BirdBeforeAdvice.java 和后置通知类 BirdAfterAdvice.java，并覆写相应方法。

【BirdBeforeAdvice.java】

```
package aop;
import org.springframework.aop.MethodBeforeAdvice;
import org.springframework.stereotype.Component;
import java.lang.reflect.Method;
@Component
public class BirdBeforeAdvice implements MethodBeforeAdvice {
    @Override
    public void before(Method method, Object[] args, Object target) throws Throwable {
        System.out.println("准备执行方法");
    }
}
```

【BirdAfterAdvice.java】

```
package aop;
import org.springframework.aop.AfterReturningAdvice;
import org.springframework.stereotype.Component;
import java.lang.reflect.Method;
@Component
public class BirdAfterAdvice implements AfterReturningAdvice {
    @Override
```

```
    public void afterReturning(Object returnValue, Method method, Object[] args,
Object target) throws Throwable {
        System.out.println("结束执行方法");
    }
}
```

在 SpringDemo3 的 src/main/java 目录下新建 config 文件夹。在 config 文件夹下新建配置类 SpringDemo3Config.java。类上添加@ComponentScan 注解并自动扫描生成 Bean，类内部配置 ProxyFactoryBean 对象并设置相关属性。

【SpringDemo3Config.java】

```
package config;
import aop.BirdInterface;
import org.springframework.aop.framework.ProxyFactoryBean;
import org.springframework.context.annotation.Bean;
import org.springframework.context.annotation.ComponentScan;
import org.springframework.context.annotation.Configuration;
@Configuration
@ComponentScan("aop")
public class SpringDemo3Config {
    //方法参数类型为被代理的目标类的类型
    @Bean
    public ProxyFactoryBean proxyFactoryBean(BirdInterface birdInterface)
throws ClassNotFoundException {
        ProxyFactoryBean factoryBean = new ProxyFactoryBean();
        //被代理的目标对象
        factoryBean.setTarget(birdInterface);
        //指定被代理的接口,若没有实现接口就采用 cglib 代理
        factoryBean.setProxyInterfaces(new Class[]{BirdInterface.class});
        //指定插入目标对象的 Bean 列表。如有多个 Bean,可以写成字符串数组
        factoryBean.setInterceptorNames(new String[]
                    {"birdBeforeAdvice","birdAfterAdvice"});
        //若设置为 true,强制使用 cglib。默认为 false
        factoryBean.setProxyTargetClass(false);
        return factoryBean;
    }
}
```

在 aop 文件夹下新建测试类 AopTest.java,在其中通过 Spring 容器获取 BirdInterface 对象，并执行 fly 方法。注意此处获取的 Bean 对象类型为接口类型,且只能通过 id 获取,写法如下:

```
BirdInterface birdInterface = (BirdInterface) ctx.getBean("proxyFactoryBean");
```

而不能使用如下写法,通过 Bean 类型获取:

```
BirdInterface birdInterface = ctx.getBean(BirdInterface.class);
```

因为 Spring 容器中有 2 个 BirdInterface 类型的 Bean 对象:一个是 Bird 对象,另一个

是 ProxyFactoryBean 代理 Bird 所生产的代理对象。

【AopTest.java】

```java
package aop;
import config.SpringDemo3Config;
import org.springframework.context.annotation.AnnotationConfigApplicationContext;
public class AopTest {
    public static void main(String [] args){
        AnnotationConfigApplicationContext ctx =
        new AnnotationConfigApplicationContext(SpringDemo3Config.class);
        BirdInterface birdInterface = (BirdInterface) ctx.getBean
("proxyFactoryBean");
        birdInterface.fly();
    }
}
```

运行结果如图 3-4 所示，可以看到前置通知和后置通知都已经插入目标方法前后了。

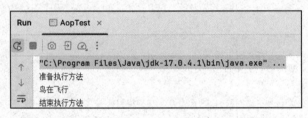

图 3-4　AopTest.java 程序的运行结果

2. 基于自定义切面类

此方法需要自己新建一个切面类，切面类需继承 MethodInterceptor 接口，并覆写 invoke 方法，然后在 Spring 配置类中利用 ProxyFactoryBean 对象配置该切面类实现 AOP。具体实现如下。

在 aop 文件夹下新建切面类 DiyAspects.java，DiyAspects 实现 MethodInterceptor 接口并覆写 invoke 方法，DiyAspects 类中定义 beforeLog 和 afterLog 来实现对目标方法的前置和后置通知。

【DiyAspects.java】

```java
package aop;
import org.aopalliance.intercept.MethodInterceptor;
import org.aopalliance.intercept.MethodInvocation;
import org.springframework.stereotype.Component;
@Component
public class DiyAspects implements MethodInterceptor {
    @Override
    public Object invoke(MethodInvocation invocation) throws Throwable {
        beforeLog();
        Object obj=invocation.proceed();
        afterLog();
        return obj;
```

```
    }
    private void beforeLog() {
        System.out.println("准备执行方法");
    }
    private void afterLog(){
        System.out.println("结束执行方法");
    }
}
```

将 SpringDemo3Config.java 类中如下代码

```
factoryBean.setInterceptorNames(new String[]{"birdBeforeAdvice","birdAfterAdvice"});
```

改成

```
factoryBean.setInterceptorNames("diyAspects");
```

运行 AopTest.java，也能看到如图 3-4 所示的正常结果。

任务 3.3 基于注解的 AOP 编程

在实际开发中，Spring 推荐开发人员使用一款功能更强大的 AOP 框架——AspectJ。AspectJ 是一个基于 Java 语言的 AOP 框架。Spring AOP 中引入了对 AspectJ 的支持，并允许直接使用 AspectJ 进行编程。

AspectJ 实现 AOP 有两种方式：一种是基于 XML，另一种是基于注解。下面主要介绍基于注解的 AspectJ 编程。

AspectJ 框架提供了一整套注解，使开发人员能够更加方便地实现 AOP 编程，AspectJ 主要注解如表 3-3 所示。

表 3-3 AspectJ 主要注解

注解名称	注 解 描 述
@Aspects	用于标识一个切面类，使用在类上
@Before	用于标识前置通知，使用在方法上。相当于 MethodBeforeAdvice，内部有一个 value 属性，value 属性可定义切入点表达式
@AfterReturning	用于标识后置通知，使用在方法上，相当于 AfterReturningAdvice，内部有两个属性为 value 和 returning。value 属性可定义一个切入点表达式；returning 属性可定义一个形参，用于在通知中获取目标方法返回值
@Around	用于标识环绕通知，使用在方法上，相当于 MethodInterceptor，内部有一个属性 value，value 属性可定义一个表达式用于指定切点
@AfterThrowing	用于标识异常处理通知，使用在方法上，相当于 ThrowsAdvice，内部有两个属性为 value 和 throwing。value 属性可定义一个切入点表达式；throwing 属性可定义一个形参，用于在通知中获取目标方法抛出的异常
@After	用于标识最终通知，使用在方法上。不管目标方法是否异常，最终都会执行，类似于异常处理中的 finally 语句

在 AspectJ 中，通常通过定义切入点表达式来指定切点。对 AspectJ 中一些常用的切入点表达式，Spring 都提供支持。其中，使用最多的切入点表达式就是 execution 切入点表达式。execution 切入点表达式的格式如下：

```
execution(modifiers-pattern? ret-type-pattern declaring-type-pattern?
    name-pattern(param-pattern) throws-pattern?)
```

(1) 修饰符匹配(modifier-pattern?)：指定目标方法的访问修饰符，如 public、private 等。

(2) 返回值匹配(ret-type-pattern)：指定目标方法的返回值类型，如 void、String 等。如果为 *，代表任意返回值。

(3) 类路径匹配(declaring-type-pattern?)：指定目标方法的类路径，如 com.my.service.UserService。如果为 *，代表所有类。

(4) 方法名匹配(name-pattern)：指定目标方法名，如 add。如果为 *，代表所有方法。

(5) 参数匹配(param-pattern)：指定具体的参数类型，多个参数间用","隔开。如(String)表示匹配一个 String 参数的方法；(,String)表示匹配有两个参数的方法，第一个参数可以是任意类型，而第二个参数是 String 类型。(…)表示任意参数。

(6) 异常类型匹配(throws-pattern?)：指定目标方法抛出的异常类型。

注意：(1)(3)(6)项后面带有"?"，在表达式中是可选项，(2)(4)(5)选项没有"?"，为必选项。各选项之间以空格分隔。

下面举例介绍一些具体的 execution 切入点表达式。

(1) 匹配所有方法，注意第一个 * 后面有空格。

```
execution(* *(..))
```

(2) 匹配 com.my.service.UserService 中所有的公有方法。

```
execution(public * com.my.service.UserService.*(..))
```

(3) 匹配 om.my.service 包下所有类的所有方法。

```
execution(* com.my.service.*.*(..))
```

(4) 匹配 UserService 类和 UserService 子类的所有方法。

```
execution(* com.my.service.UserService+.*(..))
```

(5) 匹配包中定义的所有方法，不包含子包中的方法。

```
execution(* com.my.service.*.*(..))
```

(6) 匹配包和子包中定义的所有方法。

```
exection(* com.my.service..*(..))
```

(7) 匹配以 save 开头的所有方法。

```
execution(* save*(..))
```

(8) 匹配以 save 开头且有 2 个输入参数的方法。第一个参数为 List 类型,第二个参数为 String 类型。其中,List 类型要写全类名。

```
execution(* save*(java.util.List, String))
```

(9) 匹配以 save 开头且有 2 个输入参数的方法。第一个参数为 String 类型,第二个参数可以为任意类型。

```
execution(* save*(String, *))
```

(10) 匹配以 save 开头的方法,输入参数个数不限,只需第一个参数为 String 类型。

```
execution(* save*(String, ..))
```

(11) execution 之间还可以使用逻辑运算符进行逻辑组合,如匹配 save 开头或 update 开头的所有方法。

```
execution(* save*(..))||execution(* update*(..))
```

要使用 AspectJ,首先要在 pom.xml 中引入 AspectJ 依赖。Spring 中已经集成了 AspectJ 框架,包名为 spring-aspects。只需在 SpringDemo3 项目的 pom.xml 文件下添加 Spring 6.0.3 版本的 AspectJ 依赖即可。

```
<dependency>
    <groupId>org.springframework</groupId>
    <artifactId>spring-aspects</artifactId>
    <version>6.0.3</version>
</dependency>
```

下面通过案例演示 AspectJ 注解的具体使用。

在 aop 目录下新建 MyServiceImpl.java 业务类和对应的接口类 MyService.java。MyServiceImpl 内部实现 add、delete、select、errorMethod 四个方法。add、delete、select 方法模拟业务层相关业务,errorMethod 方法中人为设置数据越界报错以用于 AfterThrowing 通知的插入。

【MyService.java】

```
package aop;
public interface MyService {
    void add();
    String delete();
    String select();
    void erroeMethod();
}
```

【MyServiceImpl.java】

```java
package aop;
import org.springframework.stereotype.Service;
@Service
public class MyServiceImpl implements MyService {
    @Override
    public void add() {
        System.out.println("添加数据");
    }
    @Override
    public String delete() {
        System.out.println("删除数据");
        String result="delete_success";
        return result;
    }
    @Override
    public String select() {
        System.out.println("查询数据");
        String result="select_success";
        return result;
    }
    @Override
    public void erroeMethod() {
        int[] a={1,2,3};
        //数组越界
        int b=a[3];
        System.out.println("更新数据");
    }
}
```

新建类 MyAnnotationAspect.java，使用@Aspects 注解标识为切面类。MyAnnotationAspect 内部定义前置通知方法 myBefore、后置通知方法 myAfter、环绕通知方法 myAround、异常处理通知方法 myAfterThrowing 和最终通知方法 myFinally，并使用相应注解标识。各通知方法内部添加代码来输出相关信息。

【MyAnnotationAspect.java】

```java
package aop;
import org.aspectj.lang.JoinPoint;
import org.aspectj.lang.ProceedingJoinPoint;
import org.aspectj.lang.annotation.*;
import org.springframework.stereotype.Component;
@Aspect
@Component
public class MyAnnotationAspect {
    //MyService 的 add 方法前置通知
    @Before("execution(* aop.MyService.add(..))")
    public void myBefore(JoinPoint joinPoint) {
```

```java
        System.out.print("目标类"+joinPoint.getTarget() );
        System.out.println("的目标方法"
                +joinPoint.getSignature().getName()+"执行前置通知");
    }
    //MyService的delete方法后置通知,并获取方法返回值
    @AfterReturning(value = "execution(* aop.MyService.delete(..))",returning = "obj")
    public void myAfter(JoinPoint joinPoint ,Object obj) {
        System.out.print("目标类"+joinPoint.getTarget() );
        System.out.println("的目标方法"
                +joinPoint.getSignature().getName()+"执行后置通知");
        System.out.println("返回值为"+obj);
    }
    /* MyService的select方法环绕通知,并获取方法返回值。注意环绕通知参数为
    ProceedingJoinPoint类型 */
    @Around("execution(* aop.MyService.select(..))")
    public Object myAround (ProceedingJoinPoint proceedingJoinPoint) throws Throwable {
        System.out.print("目标类"+proceedingJoinPoint.getTarget() );
        System.out.println("的目标方法"
                +proceedingJoinPoint.getSignature().getName()+"开始执行环绕通知");
        Object object=proceedingJoinPoint.proceed();
        System.out.println("返回值为"+object);
        System.out.print("目标类"+proceedingJoinPoint.getTarget() );
        System.out.println("的目标方法"
                +proceedingJoinPoint.getSignature().getName()+"结束执行环绕通知");
        return object;
    }
    //MyService的erroeMethod异常处理通知,并获取异常值
    @AfterThrowing (value =" execution ( *  aop. MyService. erroeMethod (..))", throwing="e")
    public void myAfterThrowing(JoinPoint joinPoint, Throwable e) {
        System.out.print("目标类"+joinPoint.getTarget() );
        System.out.println("的目标方法"
                +joinPoint.getSignature().getName()+"出错了,");
        System.out.println("错误为"+e);
    }
    //MyService所有方法的最终通知
    @After(value="execution(* aop.MyService.*(..))")
    public void myFinally(JoinPoint joinPoint) {
        System.out.print("目标类"+joinPoint.getTarget() );
        System.out.println("的目标方法"
                +joinPoint.getSignature().getName()+"执行最终通知");
    }
}
```

需要在 SpringDemo3Config.java 类上添加@EnableAspectJAutoProxy 注解,使 AspectJ 注解生效,否则所有配置的 AspectJ 注解都不会生效。注意此步非常重要。

新建测试类 MyAnnotationAspectTest.java,测试切面类中定义的五个通知方法是否正

常生效。

【MyAnnotationAspectTest.java】

```java
package aop;
import config.SpringDemo3Config;
import org.springframework.context.annotation.AnnotationConfigApplicationContext;
public class MyAnnotationAspectTest {
    public static void main(String [] args){
        AnnotationConfigApplicationContext ctx =
        new AnnotationConfigApplicationContext(SpringDemo3Config.class);
        MyService myService=(MyService)ctx.getBean("myServiceImpl");
        System.out.println("===add方法前置+最终===");
        myService.add();
        System.out.println("===delete方法后置+最终===");
        myService.delete();
        System.out.println("===select方法环绕+最终===");
        myService.select();
        System.out.println("===erroeMethod方法环绕+最终===");
        myService.erroeMethod();
    }
}
```

运行结果如图3-5所示，可以看到配置的五个通知都正常生效了。

```
===add方法前置+最终===
目标类 aop.MyServiceImpl@651aed93的目标方法 add执行前置通知
添加数据
目标类 aop.MyServiceImpl@651aed93的目标方法 add执行最终通知
===delete方法后置+最终===
删除数据
目标类 aop.MyServiceImpl@651aed93的目标方法 delete执行后置通知
返回值为 delete_success
目标类 aop.MyServiceImpl@651aed93的目标方法 delete执行最终通知
===select方法环绕+最终===
目标类 aop.MyServiceImpl@651aed93的目标方法 select开始执行环绕通知
查询数据
目标类 aop.MyServiceImpl@651aed93的目标方法 select执行最终通知
返回值为 select_success
目标类 aop.MyServiceImpl@651aed93的目标方法 select结束执行环绕通知
===erroeMethod方法环绕+最终===
目标类 aop.MyServiceImpl@651aed93,的目标方法 erroeMethod出错了，错误为
java.lang.ArrayIndexOutOfBoundsException: Index 3 out of bounds for length 3
目标类 aop.MyServiceImpl@651aed93的目标方法：erroeMethod执行最终通知
```

图3-5　MyAnnotationAspectTest.java程序的运行结果

任务3.4　综合案例：利用AOP实现访问控制

在实际应用中，AOP通常被用于对目标业务方法的执行前后添加权限判断、日志记录、事务处理等功能。下面以一个综合案例演示具体应用场景中AOP的使用。案例利用AOP实现对目标方法的访问控制。

3.4.1 案例任务

基于给定某目标方法和权限认证接口,根据用户角色进行权限认证,判断当前时间是否允许该用户访问目标方法。如果用户角色为超级管理员,则可以在任意时间访问方法;如果用户角色为普通管理员,则可以在 8:00—12:00 时间访问方法;如果用户角色为用户,则可以在 13:00—17:00 时间访问方法。访问方法时需打印出目标方法相关信息和方法运行的时间。

3.4.2 任务分析

经分析,该任务可利用 AOP 实现,编程思路为:自定义切面类,在类中注入权限认证接口对象。然后在切面类中定义环绕通知,通知内部获取目标方法输入参数值并调用权限认证接口。权限认证接口根据用户角色判断该用户是否能在当前时刻访问目标方法,若能够访问,则引出目标方法相关信息和方法运行的时间,否则拒绝访问。任务中对时间的获取与比较可使用 Java 中 LocalTime 对象的相关方法。

3.4.3 任务实施

在 aop 文件夹下新建目标类 AuthAspectsTargetClass.java,内部定义 process 方法。

【AuthAspectsTargetClass.java】

```java
package aop;
import org.springframework.stereotype.Service;
@Service
public class AuthAspectsTargetClass {
    public String process(String role) throws InterruptedException{
        System.out.println("执行目标方法");
        Thread.sleep(1000);
        return "success";
    }
}
```

在 aop 文件夹下新建权限认证接口类 AuthService.java 和接口实现类 AuthServiceImpl.java,AuthServiceImpl 内部实现 auth 方法用于权限认证。

【AuthService.java】

```java
package aop;
public interface AuthService {
    Boolean auth(String role);
}
```

【AuthServiceImpl.java】

```java
package aop;
import org.springframework.stereotype.Service;
import java.time.LocalDateTime;
```

```java
import java.time.LocalTime;
@Service
public class AuthServiceImpl implements AuthService{
    LocalTime begintime1=LocalTime.of(8,0);
    LocalTime endttime1=LocalTime.of(12,0);
    LocalTime begintime2=LocalTime.of(13,0);
    LocalTime endttime2=LocalTime.of(17,0);
    @Override
    public Boolean auth(String role) {
        LocalTime localTime=LocalTime.now();
        System.out.println("当前时间为"+localTime);
        Boolean res=false;
        //角色为超级管理员,则可以在任何时间访问方法
        if(role.equals("超级管理员")){
            res=true;
        }
        //角色为普通管理员,则可以在 8:00—12:00 时间访问方法
        else if(role.equals("普通管理员")){
            if(localTime.isAfter(begintime1)&&localTime.isBefore(endttime1)){
                res=true;
            }
        }
        //角色为用户,则可以在 13:00—17:00 时间访问方法
        else{
            if(localTime.isAfter(begintime2)&&localTime.isBefore(endttime2)){
                res=true;
            }
        }
        return res;
    }
}
```

在 aop 文件夹下新建切面类 AuthAspects.java，AuthAspects 内部实现环绕通知对 AuthAspectsTargetClass 的 process 方法进行环绕切入。

【AuthAspects.java】

```java
package aop;
import org.aspectj.lang.ProceedingJoinPoint;
import org.aspectj.lang.annotation.Around;
import org.aspectj.lang.annotation.Aspect;
import org.springframework.beans.factory.annotation.Autowired;
import org.springframework.stereotype.Component;
import java.time.Duration;
import java.time.LocalTime;
@Aspect
@Component
public class AuthAspects {
    //注入权限认证 Bean 对象
    @Autowired
```

```java
    private AuthService authService;
    //定义方法开始时间
    private LocalTime starttime;
    //定义方法结束时间
    private LocalTime endtime;
    @Around("execution( * aop.AuthAspectsTargetClass.process(..))")
    public void myAround(ProceedingJoinPoint proceedingJoinPoint) throws Throwable {
        //获取目标方法输入参数
        String role=proceedingJoinPoint.getArgs()[0].toString();
        //调用 authService 对象的 auth 方法权限认证
        boolean result=authService.auth(role);
        if(result==true){
            System.out.print("权限认证通过,目标类"
                    +proceedingJoinPoint.getTarget() );
            System.out.println("的目标方法"
                    +proceedingJoinPoint.getSignature().getName()
                    +"开始执行环绕通知");
            starttime=LocalTime.now();
            Object object=proceedingJoinPoint.proceed();
            System.out.println("方法"+proceedingJoinPoint.getSignature()
                    .getName()+"返回值为"+object.toString() );
            System.out.println("目标类"+proceedingJoinPoint.getTarget()+
                    "的目标方法"+proceedingJoinPoint.getSignature().getName()+"
                    结束执行环绕通知" );
            endtime=LocalTime.now();
            System.out.print("方法执行时间为"
                    + Duration.between(starttime,endtime).toSeconds()+"秒");
        }else{
            System.out.print("权限认证失败,用户未在指定时间访问目标方法");
        }
    }
}
```

在 aop 文件夹下新建测试类 AuthTest.java,测试权限认证功能是否正常生效。

【AuthTest.java】

```java
package aop;
import config.SpringDemo3Config;
import org.springframework.context.annotation.AnnotationConfigApplicationContext;
public class AuthTest {
    public static void main(String [] args) throws InterruptedException{
        AnnotationConfigApplicationContext ctx =
        new AnnotationConfigApplicationContext(SpringDemo3Config.class);
        AuthAspectsTargetClass authAspectsTargetClass=
        (AuthAspectsTargetClass) ctx.getBean("authAspectsTargetClass");
        authAspectsTargetClass.process("超级管理员");
    }
}
```

运行结果如图 3-6 所示,可以看到权限认证功能正常生效。当用户为角色为超级管理员,在 23:42 能够访问 process 方法。方法执行时间为 1s。可以试着将用户角色改为普通管理员,同样在 23:42 访问 process 方法,则拒绝执行。

```
当前时间为23:42:39.620067300
权限认证通过,目标类 aop.AuthAspectsTargetClass@518caac3的目标方法process开始执行环绕通知
执行目标方法
方法process返回值为success
目标类 aop.AuthAspectsTargetClass@518caac3的目标方法process结束执行环绕通知
方法执行时间为1s
```

图 3-6 AuthTest.java 程序的运行结果

小　　结

AOP 和 IOC 被称为 Spring 的两大核心思想,广泛应用代码编写中。本项目详细介绍了 Java 的代理机制、Spring AOP 的原理和具体实现、基于 AspectJ 注解的 AOP 编程等内容。最后以一个综合案例——利用 AOP 实现访问控制,介绍了开发中如何利用 AOP 实现权限认证功能,使读者对 Spring AOP 的理论和实践运用有更深入的了解。

课后练习: 利用 AOP 方法实现权限认证

定义数据操作增、删、改、查方法,利用 AOP 实现不同角色的用户执行不同方法。外部用户只能执行查询方法,业务人员只能执行增加和修改方法,超级管理员能够执行增、删、改、查所有方法。

项目 4　初识 Spring Boot

在实际开发中，Spring 应用程序的构建大多基于 Spring Boot 项目模板。Spring 官网也默认使用 Spring Boot 项目模板构建 Spring 应用程序。本章将正式进入 Spring Boot 的学习阶段，介绍的内容包括：Spring Boot 简介，如何利用 Idea 创建 Spring Boot 项目，Spring Boot 项目的结构及相关配置等内容。

任务 4.1　了解 Spring Boot

Spring Boot 是 Spring 框架中的一个项目模板。Spring Boot 使用约定优于配置的原则，使开发人员能够以最少配置创建独立的 Spring 应用程序，简化了 Spring 应用程序的配置和部署。现代的 Java 开发以微服务架构为主流，使用 Spring Boot 能够很方便地构建单体微服务项目并交给 Spring Cloud 框架统一管理。

Spring Boot 的优势包括以下方面。

（1）简化部署：Spring Boot 基于 Spring，创建的项目能以 jar 包的形式独立运行，简化了部署。

（2）简化配置：Spring Boot 提供的 starter 组件可以快速整合第三方框架，实现自动配置。

（3）简化 Web 开发：Spring Boot 集成主流 Web 开发框架，内嵌 Tomcat 等 Servlet 容器，方便 Web 开发。

（4）简化监控：Spring Boot 提供 Spring Boot Admin 等组件对微服务系统进行性能监控。

（5）简化编码：Spring Boot 能够实现完全基于注解编码，无须配置 XML。

任务 4.2　体验 Spring Boot 编程

在企业级项目开发中，为了进一步简化开发过程，一般使用 Spring Boot 构建 Spring Maven 应用程序。Spring Boot 创建的项目总体可以分为两类：一类是单体微服务项目，项目作为纯粹的服务端使用；另一类是 Web 应用，项目内嵌 Web 服务器。本任务将介绍如何使用 Idea 创建基本的 Spring Boot 项目，并对 Spring Boot 项目的项目结构和相关配置做详细介绍。

4.2.1 创建 Spring Boot 项目

Idea 默认使用 Spring Boot 项目模板创建 Spring 应用。当前与 Spring 6 匹配的 Spring Boot 稳定版为 3.0.2 版本，因此，本书统一使用 Spring Boot 3.0.2 版本来创建 Spring Boot 项目。下面演示利用 Idea 创建一个基本的 Spring Boot 项目。

创建 Spring Boot 项目

4.2.2 分析项目结构及 pom.xml 文件

Spring Boot 项目会自动帮我们加载大量依赖包和配置。4.2.1 小节中的 Spring Boot 项目创建后的结构如图 4-1 所示。Spring Boot 项目结构与 Spring 项目类型相比有一些新的变化，具体变化如下。

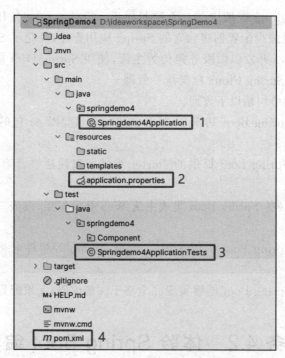

图 4-1 Spring Boot 项目结构

1. Springdemo4Application

Spring Boot 项目提供默认的项目主程序入口类 Springdemo4Application.java，类中代码如下。

```
package springdemo4;
```

```
import org.springframework.boot.SpringApplication;
import org.springframework.boot.autoconfigure.SpringBootApplication;
@SpringBootApplication
public class Springdemo4Application {
    public static void main(String[] args) {
        SpringApplication.run(Springdemo4Application.class, args);
    }
}
```

Springdemo4Application 类上添加@SpringBootApplication 注解来标识这是一个 Spring Boot 项目。被@SpringBootApplication 注解标识的类就是 Spring Boot 的主程序入口。实际上该类就是一个 Spring 配置类,也是一个 Component 组件类。Springdemo4Application 类中定义 main 方法,内部执行 SpringApplication 对象的 run 方法来运行一个 Spring 应用。run 方法格式如下:

```
SpringApplication.run(App.class, args)
```

run 方法要求传入的 App.class 必须是被@SpringBootApplication 注解标识的类。

@SpringBootApplication 是一个组合注解,用于解决 Spring Boot 主程序入口类上注解过多的问题。@SpringBootApplication 内部最重要的注解是以下三个。

(1) @SpringBootConfiguration:实现了相当于@Configuration 注解的功能,标记当前类是配置类,将配置类中定义的一个或者多个 Bean 加入 Spring 容器中。

(2) @EnableAutoConfiguration:会自动导入 Spring Boot 项目的 pom.xml 文件中引入的所有依赖包,并自动添加所需配置信息。

(3) @ComponentScan:定义自动扫描路径,在该路径下被标识的 Java 类会被自动扫描生成 Bean 并加入 Spring 容器中。默认扫描的是主程序类所在的包以及子包下的类。

2. application.properties

Spring Boot 项目提供默认的主配置文件 application.properties,配置文件内部需按照 Spring Boot 约定的配置格式定义相关配置项,这些配置项会被 Spring Boot 程序自动加载。此外,Spring Boot 也支持 yaml 格式的主配置文件。按配置文件加载优先级从高到低排列,Spring Boot 主配置文件可以放在以下 4 个位置,高优先级的配置会覆盖低优先级的配置。

优先级 1:项目根路径下的 config 文件夹。
优先级 2:项目根路径下。
优先级 3:classpath 目录下的 config 文件夹。
优先级 4:classpath 目录下。

3. Springdemo4ApplicationTests

Spring Boot 项目提供默认的项目测试类 Springdemo4Application.java,一般用于项目的单元测试,类中代码如下:

```
package springdemo4;
import org.junit.jupiter.api.Test;
import org.springframework.boot.test.context.SpringBootTest;
```

```
@SpringBootTest
class Springdemo4ApplicationTests {
    @Test
    void contextLoads() {
    }
}
```

其中,@SpringBootTest 注解标识该类为 Spring Boot 项目的测试类,@Test 是测试包 Junit 包中的注解,用于标识测试类中的测试程序入口方法。

4. pom.xml

pom.xml 文件中默认引入了一个父项目 spring-boot-starter-parent。由于创建项目时勾选了 Spring Web 模块,因此,pom.xml 文件中还引入了 spring-boot-starter-web、spring-boot-starter-test 和 spring-boot-maven-plugin 三个依赖包,代码如下。

```xml
<?xml version="1.0" encoding="UTF-8"?>
<project xmlns="http://maven.apache.org/POM/4.0.0"
    xmlns:xsi="http://www.w3.org/2001/XMLSchema-instance"
        xsi:schemaLocation="http://maven.apache.org/POM/4.0.0
                   https://maven.apache.org/xsd/maven-4.0.0.xsd">
    <modelVersion>4.0.0</modelVersion>
    <parent>
        <groupId>org.springframework.boot</groupId>
        <artifactId>spring-boot-starter-parent</artifactId>
        <version>3.0.2</version>
        <relativePath/> <!-- lookup parent from repository -->
    </parent>
    <groupId>springdemo4</groupId>
    <artifactId>springdemo4</artifactId>
    <version>0.0.1-SNAPSHOT</version>
    <name>springdemo4</name>
    <description>Demo project for Spring Boot</description>
    <properties>
        <java.version>17</java.version>
    </properties>
    <dependencies>
        <dependency>
            <groupId>org.springframework.boot</groupId>
            <artifactId>spring-boot-starter-web</artifactId>
        </dependency>
        <dependency>
            <groupId>org.springframework.boot</groupId>
            <artifactId>spring-boot-starter-test</artifactId>
            <scope>test</scope>
        </dependency>
    </dependencies>
    <build>
        <plugins>
            <plugin>
```

```
            <groupId>org.springframework.boot</groupId>
            <artifactId>spring-boot-maven-plugin</artifactId>
        </plugin>
     </plugins>
   </build>
</project>
```

（1）spring-boot-starter-parent：该父项目有两方面功能：一方面用于统一指定项目配置，如指定项目编码格式、指定 Java 版本、指定 Spring Boot 版本等；另一方面用于统一管理各依赖包版本，在引入依赖包时开发人员可以不指定版本，Spring Boot 帮我们自动指定匹配的版本。

（2）spring-boot-starter-web：这是一个 Spring Boot 项目的 Web 启动器，用于指定在 Spring Boot 启动时导入 Web 开发所需的依赖。在 Spring Boot3 中，starter-web 内部直接嵌入了新版本的 Servlet 容器（默认为 Tomcat 10），自动加载 Spring MVC 相关配置实现 HTTP 接口，简化了 Web 开发。

（3）spring-boot-starter-test：spring-boot-starter-test 中默认引入 Junit 单元测试包，项目中不需手动引入 Junit 也能进行单元测试。

（4）spring-boot-maven-plugin：这是 maven 为 Spring Boot 项目提供的项目构建插件包，在 Spring Boot3 中该插件包内部通过 SpringApplicationAotProcessor 类引入了预编译机制，使 Spring Boot 项目在编译阶段完成 Bean 的扫描和注册，加快项目的启动速度。

Spring Boot 将应用开发中所有的功能都变成了一个个启动器，类似于 spring-boot-starter-web，通过 spring-boot-starter 这种统一格式导入。

4.2.3 运行 Spring Boot 项目并打包

Spring Boot 项目创建完毕，则直接运行主程序入口类 Springdemo4Application，项目就可以正常启动了，在控制台可以看见如下日志信息。

```
Tomcat started on port(s): 8080 (http) with context path ''
```

由于我们之前创建的项目引入了 spring-boot-starter-web，项目启动后会同时开启 Tomcat 的 8080 端口服务，成为一个 Web 应用。这时浏览器访问项目地址 http://localhost:8080。看到界面有如下信息，因为 Spring Boot 项目已经启动成功且默认没有创建任何页面，所以浏览器会访问 error 接口。

```
Whitelabel Error Page
This application has no explicit mapping for /error, so you are seeing this as a
fallback.
Sun Feb 26 11:49:20 CST 2023
There was an unexpected error (type=Not Found, status=404).
```

下面在 SpringDemo4 项目内添加一个简单的功能以演示项目打包效果。在 springdemo4 包下新建目录 component，在 component 目录下新建一个接口类 HelloWorldController.java，内部定义一个 HelloWorld 接口方法，对外返回"helloworld"字符串。

【HelloWorldController.java】

```
package springdemo4.Component;
import org.springframework.web.bind.annotation.GetMapping;
import org.springframework.web.bind.annotation.RestController;
@RestController
public class HelloWorldController {
    @GetMapping("/helloWorld")
    public String helloWorld(){
        return "helloworld";
    }
}
```

其中，@RestController 注解和 @GetMapping 注解是 Spring MVC 中的两个注解。@RestController 注解用于将 HelloWorldController 类标识成一个控制层类，该类所有的方法返回值都将输出到页面。@GetMapping 注解用于标识 helloWorld 方法为一个 Web 请求的映射接口方法。这两个注解将在后续内容中进行详细介绍。

利用 Maven 将 SpringDemo4 项目打包成 Jar，直接部署在 JVM 上运行。在 Idea 右侧边栏单击 Maven 选项，弹出如图 4-2 所示界面。在界面中双击 package 选项，开始为项目打包。打包结束后，如果控制台看到 BUILD SUCCESS 信息，则打包成功。生成的 Jar 包在 target 目录下，名字为 springdemo4-0.0.1-SNAPSHOT.jar。

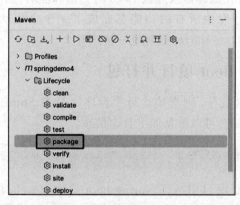

图 4-2 打包 Spring Boot 项目

在 Windows 中打开 target 目录，在 target 目录的地址栏中输入 cmd 命令，在当前路径下打开 Windows 命令行工具，在 Windows 命令行工具输入命令来运行 Jar。

```
java -jar springdemo4-0.0.1-SNAPSHOT.jar
```

等待 Jar 包启动完毕，在浏览器中输入访问地址 http://localhost:8080/helloWorld。可以看到页面显示 helloworld 字符串。此时 Spring Boot 项目就相当于一个服务的后端，对外提供 API 接口，实现了 Web 开发的前后端分离。因此，我们可以把一个复杂的业务拆分为一个个模块，每个模块利用 Spring Boot 构建单体的微服务项目，实现业务模块化。

在 Spring Boot 项目中引入 spring-boot-starter-web 创建的就是一个 Web 项目，Spring Boot

启动默认会开启 Tomcat 服务。spring-boot-starter-web 默认会引入 Spring MVC 框架,Spring MVC 相关知识将在后续章节中介绍,下面仍然以单体微服务形式来介绍 Spring Boot。

4.2.4 设置 Spring Boot 服务开机启动

一般在利用 Spring Boot 构建单体微服务时,需要在项目内部编写自己的服务,让服务开机启动。实现此需求常见的方法有 5 种,如表 4-1 所示。

表 4-1 Spring Boot 服务开机的 5 种方法

方　　法	使 用 方 法
@PostConstruct 注解	自定义启动服务类,添加@Component 注解。类中定义方法添加服务代码,方法用@PostConstruct 注解修饰,此时服务代码将在执行启动类构造方法后执行,但属性不一定都完成注入
InitializingBean 接口	自定义启动服务类,添加@Component 注解。类实现了 InitializingBean 接口,内部覆写 afterPropertiesSet 方法并添加服务代码,此时服务代码将在启动类执行构造方法且属性注入完后执行,执行顺序晚于@PostConstruct
ApplicationRunner 接口	自定义启动服务类,添加@Component 注解。类实现了 ApplicationRunner 接口,内部覆写 run 方法并添加服务代码,此时服务代码将在 Spring 应用启动成功后执行
CommandLineRunner 接口	自定义启动服务类,添加@Component 注解。类实现了 CommandLineRunner 接口,内部覆写 run 方法并添加服务代码,此时服务代码将在 Spring 应用启动成功后执行
ApplicationListener 接口	自定义启动服务类,添加@Component 注解。类实现了 ApplicationListener 接口,接口监听事件类型为 ApplicationStartedEvent 或 ApplicationReadyEvent,接口内部覆写 onApplicationEvent 方法并添加服务代码,此时服务代码将在 Spring Boot 应用启动完毕执行

下面分别演示这 5 种方法的具体使用。在 springdemo4 包下新建一个服务类 MyApplicationStartService.java,类中编写如下代码。

【MyApplicationStartService.java】

```
package springdemo4.Component;
import jakarta.annotation.PostConstruct;
import org.springframework.beans.factory.InitializingBean;
import org.springframework.boot.ApplicationArguments;
import org.springframework.boot.ApplicationRunner;
import org.springframework.boot.CommandLineRunner;
import org.springframework.boot.context.event.ApplicationStartedEvent;
import org.springframework.context.ApplicationListener;
import org.springframework.stereotype.Component;
@Component
public class MyApplicationStartService implements InitializingBean,
ApplicationListener<ApplicationStartedEvent> , ApplicationRunner,
CommandLineRunner {
    public MyApplicationStartService(){
        System.out.println("MyApplicationStartServiceg 构造函数执行");
    }
    @Override
```

```java
    public void afterPropertiesSet() throws Exception {
        System.out.println("InitializingBean 执行");
    }
    @PostConstruct
    public void postMethod(){
        System.out.println("PostConstruct 执行");
    }
    @Override
    public void run(ApplicationArguments args) throws Exception {
        System.out.println("ApplicationRunner 执行");
    }
    @Override
    public void run(String... args) throws Exception {
        System.out.println("CommandLineRunner 执行");
    }
    @Override
    public void onApplicationEvent(ApplicationStartedEvent event) {
        System.out.println("ApplicationListener 执行");
    }
}
```

上述代码在 MyApplicationStartService 类上实现 InitializingBean、ApplicationRunner、CommandLineRunner、ApplicationListener 4 个接口,其中 ApplicationListener 监听 ApplicationStartedEvent 事件。类内部覆写 4 个接口相关方法并添加打印输出语句,再定义 postmethod 方法并用@PostConstruct 注解标识。

运行主程序入口类 Springdemo4Application,可以看到控制台有如下输出。

```
MyApplicationStartServiceg 构造函数执行
PostConstruct 执行
InitializingBean 执行
ApplicationListener 执行
ApplicationRunner 执行
CommandLineRunner 执行
```

如果把 ApplicationListener 监听事件改为 ApplicationReadyEvent,则控制台输出如下结果。

```
MyApplicationStartServiceg 构造函数执行
PostConstruct 执行
InitializingBean 执行
ApplicationRunner 执行
CommandLineRunner 执行
ApplicationListener 执行
```

从中可以得出代码的执行顺序。

(1) 构造函数执行完毕,先执行@PostConstruct 注解所在方法,然后执行 InitializingBean 的 afterPropertiesSet 方法。

(2) ApplicationListener 如果监听的是 ApplicationStartedEvent 事件,ApplicationListener 的 onApplicationEvent 方法先执行,随后执行 ApplicationRunner 的 run 方法,最后执行

CommandLineRunner 的 run 方法。

（3）ApplicationListener 如果监听的是 ApplicationReadyEvent 事件，先执行 ApplicationRunner 的 run 方法，然后执行 CommandLineRunner 的 run 方法，最后执行 ApplicationListener 的 onApplicationEvent 方法。

（4）ApplicationRunner 的 run 方法默认比 CommandLineRunner 的 run 方法先执行。

如果需要编写多个服务启动类自定义服务启动顺序，可以在服务启动类使用@Order 注解，@Order 注解按照大小顺序执行，数字越小则优先级越高。@Order 使用方法如下。

```
@Component
@Order(value = 1)
public class applicationRunner1 implements ApplicationRunner{
    @Override
    public void run(ApplicationArguments args) throws Exception {
        System.out.println("服务第一个启动");
    }
}
@Component
@Order(value = 2)
public class applicationRunner2 implements ApplicationRunner{
    @Override
    public void run(ApplicationArguments  args) throws Exception {
        System.out.println("服务第二个启动");
    }
}
```

任务 4.3　体验 Spring Boot 单元测试

在项目开发中一般都需要进行单元测试。Spring Boot 封装了单元测试组件 spring-boot-starter-test，内部使用 JUnit 进行单元测试。

4.3.1　使用默认测试类进行单元测试

在 SpringDemo4 项目中，test 文件夹下提供了默认的单元测试类 Springdemo4-ApplicationTests.java。Springdemo 4 ApplicationTests 被@SpringBootTest 注解标识为 Spring Boot 项目测试类，会在测试时自动加载@SpringBootApplication 或@SpringBootConfiguration 注解标识的相关配置，构建测试依赖环境。@Test 注解标识测试类运行时的入口方法为 contextLoads 方法。

使用 Spring Boot 单元测试非常简单，我们可以直接在 contextLoads 方法中输入如下功能代码，如打印输出"hello world"。

```
@Test
void contextLoads() {
    System.out.println("hello world");
}
```

直接运行 Springdemo4ApplicationTests 类,看到如图 4-3 所示结果。表明测试运行成功。

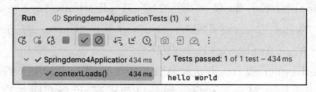

图 4-3 测试结果

4.3.2 手动创建测试类进行单元测试

在开发中,如果不想使用 Spring Boot 默认的测试类进行测试,也可以手动创建测试类。下面演示如何手动创建测试类进行服务的单元测试。

在 SpringDemo4 下的 component 目录新建 User 类,User 类使用@Component 标注为一个组件类 Bean,内部定义 print 方法返回 hellworld。User 类内部代码如下。

【User.java】

```
package springdemo4.Component;
import org.springframework.stereotype.Component;
@Component
public class User {
    public String print(){
        return "helloworld";
    }
}
```

在 User 类中右击,在弹出的菜单中选中 Go To 选项,弹出右侧子菜单,在子菜单中单击 Test 选项,如图 4-4 所示。

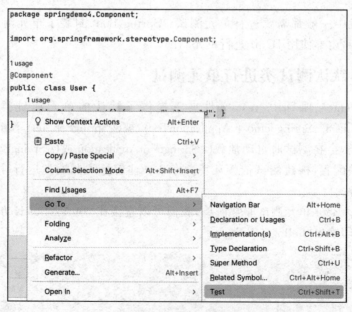

图 4-4 创建测试类 1

在弹出的界面中单击"Create New Test...."选项,进入创建测试类界面,如图 4-5 所示。页面显示测试工具默认为 Junit5。

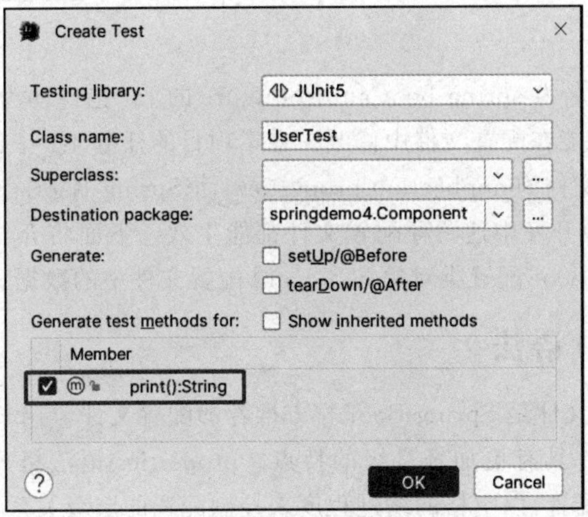

图 4-5　创建测试类 2

在界面中勾选 print 方法作为测试方法。单击 OK 按钮,在 test 文件下就会生成一个和 User 相同层级目录的测试文件 UserTest。在 UserTest 文件上添加@SpringBootTest 注解,并注入需要测试的 User 对象,然后在单元测试中调用 print 方法。

```
package springdemo4.Component;
import org.junit.jupiter.api.Test;
import org.springframework.beans.factory.annotation.Autowired;
import org.springframework.boot.test.context.SpringBootTest;
@SpringBootTest
class UserTest {
    @Autowired
    private User user;
    @Test
    void print() {
        System.out.println(user.print());
    }
}
```

运行 UserTest 类,测试结果如图 4-6 所示。

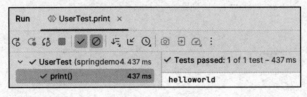

图 4-6　测试结果

任务 4.4　了解 Spring Boot 配置文件

Spring Boot 通过名为 spring-boot-autoconfigure 的 Jar 包自动为项目进行了许多默认配置，这些默认配置也能在配置文件中修改。在本项目的任务 4.2 中创建的 Spring Boot 项目默认生成的主配置文件为 application.properties，而 Spring Boot 官方推荐的主配置文件是 application.yaml，在开发中这两种配置文件都能生效。下面将介绍 yaml 配置文件的语法以及如何向 Spring Boot 配置类对象注入 yaml 配置文件中的数据。

4.4.1　初识 yaml 语法

yaml 格式的配置文件是 Spring Boot 官方推荐的配置文件，与 properties 和 XML 等配置文件相比，yaml 文件具有更加轻量级的特点。properties 语法格式为 key=value，yaml 使用键值对形式定义配置，以空格的缩进程度来控制配置的层级关系，语法格式为"key:（空格）value"。例如配置服务端口号，application.properties 文件定义如下。

```
server.port=8080
```

而 yaml 文件定义如下。

```
server:
  port: 8080
```

其中，空格缩进代表层级关系，必须严格控制。这种方式使配置文件的结构更加清晰，易于理解。yaml 支持普通键值对、对象、集合三种数据结构，也支持复合结构。下面结合一些具体例子演示 yaml 格式的配置。

```
#字符串
name: zhangsan
#整型
age: 10
#日期 Date
birthday: 2023/03/03
#person 对象
person:
  name: 10
# map 集合
map: "{name:'aa',age:10}"
# list 集合
list1:
  aa
  bb
  cc
list2: aa,bb,cc
```

yaml 语法中的注意事项总结如下。
(1) 对大小写敏感。
(2) 使用缩进表示层级关系。
(3) 不能使用 Tab 键缩进,只能使用空格键缩进。
(4) 缩进长度没有限制,只要元素对齐,就表示这些元素属于一个层级。
(5) 如需注释,使用#。
(6) 字符串可以不加引号。

4.4.2 读取 yaml 中的配置

Spring Boot 默认读取名为 application.yaml 或 application.properties 的主配置文件加载配置。这里以 application.yaml 为例,假设已经在 yaml 配置文件中自定义了一些配置,如何将 yaml 中配置的数据注入 Spring Boot 配置类对象? 比较常用的方法有 @Value 注解和@ConfigurationProperties 注解两种。其中,@Value 注解用于对单个属性注入,@ConfigurationProperties 注解用于批量注入。

1. @Value

2.2.3 小节中我们提到@Value 注解可以用于为 Bean 的属性注入外部配置文件数据,一般用于对 Bean 的每个属性单独注入,不支持批量注入。使用时,@Value 注解要求属性名和配置文件 key 值严格对应,一般用于简单数据结构的注入。下面演示如何使用@Value 注解注入 yaml 格式数据。定义如下的 application.yaml 配置文件。

【application.yaml】

```
#字符串
name: zhangsan
#整型
age: 10
#Date
birthday: 2023/03/03
#带有层级关系键值对:person.name
person:
  name: 10
```

其中,name 值是字符串,age 为整型,birthday 为日期型,person.name 为层级关系键值对。

在 component 文件夹下创建如下 DiyConfigWithValue 类,用于注入配置文件中的数据。DiyConfigWithValue 类内部可以定义属性和相应的 toString 方法。

【DiyConfigWithValue.java】

```
package springdemo4.Component;
import org.springframework.beans.factory.annotation.Value;
import org.springframework.stereotype.Component;
import java.util.Date;
import java.util.List;
import java.util.Map;
@Component
public class DiyConfigWithValue {
```

```
    @Value("${name}")
    private String name;
    @Value("${age}")
    private Integer age;
    @Value("${birthday}")
    private Date birthday;
    @Value("${person.name}")
    private String person;
    @Override
    public String toString() {
        return "DiyConfigWithValue{" +
                "name='" + name + '\'' +
                ", age=" + age +
                ", birthday=" + birthday +
                ", person='" + person + '\'' +
                '}';
    }
}
```

在 Spring Boot 默认测试类 Springdemo4ApplicationTests 中添加如下粗体测试代码，将 DiyConfigWithValue 对象注入测试类并打印 DiyConfigWithValue 对象属性内容。

```
package springdemo4;
import org.junit.jupiter.api.Test;
import org.springframework.beans.factory.annotation.Autowired;
import org.springframework.boot.test.context.SpringBootTest;
import springdemo4.Component.DiyConfigWithValue;
@SpringBootTest
class Springdemo4ApplicationTests {
    @Autowired
    private DiyConfigWithValue diyConfigWithValue;
    @Test
    void contextLoads() {
        System.out.println(diyConfigWithValue);
    }
}
```

运行测试类 Springdemo4ApplicationTests，控制台输出如下结果。可以看到配置文件数据成功注入 DiyConfigWithValue 类的所有属性中。

```
DiyConfigWithValue{name='zhangsan', age=10, birthday=Fri Mar 03 00:00:00 CST 2023, person='10'}
```

2. @ConfigurationProperties

在开发中，如果数据结构复杂，有大量自定义对象存在，就不适合用@Value 注入值了。这时就可以使用@ConfigurationProperties。@ConfigurationProperties 支持复杂数据类型的批量注入，比@Value 使用起来更加方便。使用@ConfigurationProperties 时，只需将所有的数据统一写在 yaml 的某个层级下,如下配置都写在 diyconfig 层级下。

```yaml
diyconfig:
  name: zhangsan
  age: 10
```

然后在需要注入配置数据的 Java 类上使用@ConfigurationProperties 注解,示例如下。

```
@ConfigurationProperties(prefix="diyconfig")
```

Spring Boot 会自动读取配置文件 diyconfig 层级下的所有数据,按照键名匹配属性名的规则批量注入值,如 name 的值 zhangsan 就注入 java 类中的 name 属性。

下面演示@ConfigurationProperties 的具体使用。为了与@value 区分配置文件,这里单独新建一个配置文件 application-prefix.yaml,在其中定义如下配置项进行注入,配置项既包含简单数据结构,也包含复杂数据结构,如 List、Map 和自定义对象等。

【application-prefix.yaml】

```yaml
diyconfig:
  #字符串
  name: zhangsan
  #整型
  age: 10
  #Date
  birthday: 2023/03/03
  #Person 对象
  person:
    name: lisi
    age: 20
  #Map
  map:
    k1: v1
    k2: v2
  #Person 对象集合
  list:
    - name: zhangsan1
      age: 1
    - name: zhangsan2
      age: 2
```

在 component 文件夹下新建 Person 类,内部定义 name 属性和 age 属性,并添加 get/set 方法,覆写 toString 方法。Person 类添加@Component 注解标识为一个组件类 Bean。

【Person.java】

```java
package springdemo4.Component;
import org.springframework.stereotype.Component;
@Component
public class Person {
    private String name;
```

```
    private Integer age;
    public String getName() {
        return name;
    }
    public void setName(String name) {
        this.name = name;
    }
    public Integer getAge() {
        return age;
    }
    public void setAge(Integer age) {
        this.age = age;
    }
    @Override
    public String toString() {
        return "Person{" +
            "name='" + name + '\'' +
            ", age=" + age +
            '}';
    }
}
```

在 component 文件夹下新建 DiyConfigWithPrefix 类，在 DiyConfigWithPrefix 类中定义不同类型的属性，用于注入配置文件数据。属性类型包括 String、Int、Date、自定义 Person 对象、Map 和 List<Person>对象集合，并覆写 DiyConfigWithPrefix 类的 toString 方法。DiyConfigWithPrefix 类利用@Component 标识为组件类 Bean，同时在类上添加注解：

```
@PropertySource(value = {"classpath:application-prefix.yaml"})
```

用于加载配置文件 application-prefix.yaml，添加注解：

```
@ConfigurationProperties(prefix="diyconfig")
```

用于读取 application-prefix.yaml 配置文件中 diyconfig 层级所属配置。DiyConfigWithPrefix 类代码如下。

【DiyConfigWithPrefix.java】

```
package springdemo4.Component;
import org.springframework.boot.context.properties.ConfigurationProperties;
import org.springframework.context.annotation.PropertySource;
import org.springframework.stereotype.Component;
import java.util.Date;
import java.util.List;
import java.util.Map;
@Component
@ConfigurationProperties(prefix = "diyconfig")
@PropertySource(value = {"classpath:application-prefix.yaml"})
```

```java
public class DiyConfigWithPrefix {
    private String name;
    private Integer age;
    private Date birthday;
    private Person person;
    private Map<String,Object> map;
    private List<Person> list;
    public String getName() {
        return name;
    }
    public void setName(String name) {
        this.name = name;
    }
    public Integer getAge() {
        return age;
    }
    public void setAge(Integer age) {
        this.age = age;
    }
    public Date getBirthday() {
        return birthday;
    }
    public void setBirthday(Date birthday) {
        this.birthday = birthday;
    }
    public Person getPerson() {
        return person;
    }
    public void setPerson(Person person) {
        this.person = person;
    }
    public Map<String, Object> getMap() {
        return map;
    }
    public void setMap(Map<String, Object> map) {
        this.map = map;
    }
    public List<Person> getList() {
        return list;
    }
    public void setList(List<Person> list) {
        this.list = list;
    }
    @Override
    public String toString() {
        return "DiyConfig{" +
                "name='" + name + '\'' +
                ", age=" + age +
                ", birthday=" + birthday +"\n"+
                ", person=" + person +
```

```
                ", map=" + map +"\n"+
                ", list=" + list +
                '}';
    }
}
```

引入@ConfigurationProperties 注解以后，Idea 上一般会出现如下提示信息。

```
Spring Boot Configuration Annotation Processor not found in claspath
```

这并不影响代码运行，可以忽略。如想消除，只需在 pom.xml 文件中引入以下依赖，然后重启 Idea。

```xml
<dependency>
    <groupId>org.springframework.boot</groupId>
    <artifactId>spring-boot-configuration-processor</artifactId>
    <optional>true</optional>
</dependency>
```

此时如果直接运行 Springdemo4ApplicationTests 测试类，是无法读取到 application-prefix.yaml 配置文件数据的，所有属性注入值全是 null。Spring Boot 默认读取名为 application.yaml 或 application.properties 的主配置文件加载配置。如果要读取自定义配置文件 application-prefix.yaml，可以在 application.yaml 文件中添加如下配置，激活并加载 application-prefix.yaml 配置文件。

```yaml
#加载名为 application-prefix.yaml 的配置文件
spring:
  profiles:
    active: prefix
```

在 Spring Boot 默认测试类 Springdemo4ApplicationTests 中添加如下黑体测试代码，将 DiyConfigwithValue 对象注入测试类。

【Springdemo4ApplicationTests.java】

```java
package springdemo4;
import org.junit.jupiter.api.Test;
import org.springframework.beans.factory.annotation.Autowired;
import org.springframework.boot.test.context.SpringBootTest;
import springdemo4.Component.DiyConfigWithPrefix;
@SpringBootTest
class Springdemo4ApplicationTests {
    @Autowired
    private DiyConfigWithPrefix diyConfigWithPrefix;
    @Test
    void contextLoads() {
        System.out.println(diyConfigWithPrefix);
    }
}
```

运行测试类 Springdemo4ApplicationTests,控制台输出如下结果。可以看到配置文件数据成功注入 DiyConfigWithPrefix 对象的所有属性中。

```
DiyConfig{name='zhangsan', age=10, birthday=Fri Mar 03 00:00:00 CST 2023,
person=Person{name='lisi', age=20}, map={k1=v1, k2=v2},
list=[Person{name='zhangsan1', age=1}, Person{name='zhangsan2', age=2}]}
```

在实际应用中,如果注入类中单个属性,可以使用@Value。如果注入一个复杂的 JavaBean 对象,推荐使用@ConfigurationProperties。在配置时需注意 Spring Boot 的各配置项名称不能随意更改,都必须使用约定的配置项名称,否则配置不能自动注入生效,这也是 Spring Boot 约定大于配置思想的体现。本书后续所有项目的配置文件统一采用 application.yaml 格式。

任务 4.5　Spring Boot 多环境配置

在项目开发中,通常会将项目运行环境划分为开发环境、测试环境、生产环境等,各环境的配置是不同的。Spring Boot 提供了多环境配置文件切换功能,方便开发人员在各个环境中切换配置。多环境配置有两种具体的实现方式:一种是基于多文件的多环境配置,另一种是基于单文件的多环境配置。下面将分别介绍。

4.5.1　基于多文件的多环境配置

这里以 yaml 格式配置文件为例,基于多文件的多环境配置指的是项目多环境配置分别写在不同文件中。如 application-dev.yaml 表示开发环境配置,application-test.yaml 表示测试环境配置,application-prod.yaml 表示生产环境配置。Spring Boot 启动默认不会直接加载这些配置文件,它默认还是加载 application.yaml 主配置文件。但是 Spring Boot 提供以下配置项。

```
spring:
  profiles:
    active:
```

允许我们在主配置文件中配置,使 Spring Boot 在启动时能够加载文件名是 application-{profile}.yaml 配置文件,来切换不同配置。

假设有如下三个配置文件,application-dev.yml 对应开发环境配置,其中端口为 8081。application-test.yaml 对应测试环境配置,其中端口为 8082,application-prod.yaml 对应生产环境配置,其中端口为 8083。

【application-dev.yaml】

```
server:
  port: 8081
```

【application-test.yaml】

```yaml
server:
  port: 8083
```

【application-prod.yaml】

```yaml
server:
  port: 8083
```

如果要加载生产环境配置,可以在原有的 appliaction.yaml 主配置文件中,改动如下粗体代码。

```yaml
#启动加载 application-dev.yaml
spring:
  profiles:
    active: prefix,dev
```

启动 Spring Boot 项目,即可看到控制台输出中包含如下 tomcat 启动端口为 8081 的信息,证明 application-dev.yaml 配置生效。如果要加载测试环境配置,只需把 appliaction.yaml 中的 active: dev 改成 active: test 即可。加载生产环境配置也类似。

4.5.2 基于单文件的多环境配置

除了多文件的多环境配置,yaml 还支持在单个配置文件中以文档快的形式来实现多环境配置。在 yaml 中可以直接以三个横杠来区分不同的配置。假设要加载开发环境配置,可以在主配置文件采用如下粗体写法。

```yaml
spring:
  profiles:
    active: prefix,dev
# 开发环境配置
---
spring:
  config:
    activate:
      on-profile: dev
server:
  port: 8081
---
# 测试环境配置
---
spring:
  config:
    activate:
      on-profile: test
server:
  port: 8082
```

```yaml
---
# 生产环境配置
---
spring:
  config:
    activate:
      on-profile: prod
server:
  port: 8083
---
```

启动 Spring Boot 项目,也可看到 4.5.1 小节中控制台的输出。如果同时在 application.yaml 和 application-dev.yaml 中都设置了 dev 开发环境配置项,则 application-dev.yaml 中的配置项生效。

任务 4.6 综合案例:用 Spring Boot 实现基于 TCP 服务的请求响应

开发中,Spring Boot 经常被用来编写微服务后台。下面以一个综合案例演示如何利用 Spring Boot 编写后台微服务并与客户端进行简单交互。

4.6.1 案例任务

编码实现一个基于 TCP Socket 的 Spring Boot 服务端。在 Spring Boot 启动后,程序自动加载自定义配置文件中配置项并创建 Socket 服务端。服务端能够接收客户端请求并返回响应。如客户端发送 get server config 请求用于获取服务端配置,服务端收到请求后,需将 Socket 服务端配置信息发送给客户端。

4.6.2 任务分析

该任务涉及 Spring Boot 自定义服务的开机启动,Spring Boot 自定义配置文件的加载以及 Bean 的依赖注入、Socket 编程等内容。编程思路如下。

(1) 创建自定义配置文件,配置文件内部定义若干 Socket 服务端配置项。
(2) 创建 Java 配置类,用于注入配置文件中的 Socket 服务端配置。
(3) 利用 4.2.4 小节中的方法在 Spring Boot 启动后创建一个 Socket 服务端,Socket 服务端配置由 Java 配置类传入。
(4) 编写 Socket 服务端收发数据代码。Socket 服务端接收到 get server config 请求后,将返回 Socket 服务端配置信息给客户端。
(5) 编写 Socket 客户端代码。
(6) 启动服务端和客户端进行数据交互,验证结果。

4.6.3 任务实施

在 resource 目录下创建自定义配置文件 application-tcpservercfg.yaml，其中写入如下 Socket 配置项，分别对 Socket 服务端端口和地址是否复用及接收端最大缓存进行配置。

【application-tcpservercfg.yaml】

```yaml
tcpserverconfig:
  port: 8085                    #端口
  reuseAddress: true            #地址复用
  receiveBufferSize: 2048       #接收端最大缓存
```

在 Component 目录下新建 TcpServerConfig 类，用于注入 application-tcpservercfg.yaml 中的配置数据。

【TcpServerConfig.java】

```java
package springdemo4.Component;
import org.springframework.boot.context.properties.ConfigurationProperties;
import org.springframework.context.annotation.PropertySource;
import org.springframework.stereotype.Component;
@Component
@ConfigurationProperties(prefix = "tcpserverconfig")
@PropertySource(value = {"classpath:application-tcpservercfg.yaml"})
public class TcpServerConfig {
    private Integer port;
    private boolean reuseAddress;
    private Integer receiveBufferSize;
    public Integer getPort() {
        return port;
    }
    public void setPort(Integer port) {
        this.port = port;
    }
    public boolean isReuseAddress() {
        return reuseAddress;
    }
    public void setReuseAddress(boolean reuseAddress) {
        this.reuseAddress = reuseAddress;
    }
    public Integer getReceiveBufferSize() {
        return receiveBufferSize;
    }
    public void setReceiveBufferSize(Integer receiveBufferSize) {
        this.receiveBufferSize = receiveBufferSize;
    }
}
```

创建服务启动监听类 TcpApplicationListener 并实现 ApplicationListener 接口。类内部覆写 onApplicationEvent 方法，在方法内定义一个线程来创建 Socket 服务端。

【TcpApplicationListener.java】

```java
package springdemo4.Component;
import org.springframework.beans.factory.annotation.Autowired;
import org.springframework.boot.context.event.ApplicationReadyEvent;
import org.springframework.context.ApplicationListener;
import org.springframework.stereotype.Component;
import java.io.IOException;
@Component
public class TcpApplicationListener implements
        ApplicationListener<ApplicationReadyEvent> {
    @Autowired
    private TcpServerConfig tcpServerConfig;
    @Override
    public void onApplicationEvent(ApplicationReadyEvent event) {
        try{
            //新建一个线程,创建 Socket 服务端
            new Thread(new TcpServer(tcpServerConfig)).start();
        }catch(IOException e){
            e.printStackTrace();
        }
    }
}
```

创建 TcpServer 类并实现 Runnable 接口。类内部通过构造函数创建 ServerSocket 对象并加载相关配置。类内部覆写 run 方法阻塞,等待客户端请求,如收到请求,则开启一个线程处理;请求处理完毕,将服务端相关配置发送给客户端。

【TcpServer.java】

```java
package springdemo4.Component;
import org.springframework.stereotype.Component;
import java.io.IOException;
import java.io.InputStream;
import java.io.OutputStream;
import java.net.ServerSocket;
import java.net.Socket;
public class TcpServer implements Runnable {
    private ServerSocket serverSocket=null;
    //构造函数中创建 ServerSocket 服务端对象并加载配置
    public TcpServer (TcpServerConfig tcpServerConfig) throws IOException {
        serverSocket=new ServerSocket(tcpServerConfig.getPort());
        serverSocket.setReuseAddress(tcpServerConfig.isReuseAddress());
        serverSocket.setReceiveBufferSize(tcpServerConfig.getReceiveBufferSize());
    }
    @Override
    public void run()  {
        try{
            while(true){
                //阻塞等待客户端请求,收到请求则开启一个线程处理
                Socket socket = serverSocket.accept();
                Thread thread = new Thread(){
                    public void run(){
```

```java
                        //处理请求并返回响应
                        processData(socket);
                    }
                };
                thread.start();
            }
        }catch(IOException e){
            e.printStackTrace();
        }
    }
    public void processData(Socket socket) {
        try{
            //获取输入流并读数据
            InputStream inputStream = socket.getInputStream();
            byte[] bytes = new byte[1024];
            int len = inputStream.read(bytes);
            String data = new String(bytes,0,len);
            System.out.println("收到客户端数据:"+data);
            int port=serverSocket.getLocalPort();
            Boolean reuseAddress=serverSocket.getReuseAddress();
            int receiveBufferSize=socket.getReceiveBufferSize();
            String sendMsg="TCP 服务器端口为"+port+",地址是否复用为"+
                    reuseAddress+",接收端缓存大小为"+receiveBufferSize;
            //获取输出流写数据
            OutputStream outputStream = socket.getOutputStream();
            outputStream.write(sendMsg.getBytes());
        }
        catch(IOException e){
            e.printStackTrace();
        }
    }
}
```

创建 Socket 客户端 TcpClient 类。类中定义 main 方法,在 main 方法中创建 Socket 客户端对象,利用 Socket 客户端对象发送请求和接收服务端响应。

【TcpClient.java】

```java
package springdemo4.Component;
import java.io.IOException;
import java.io.InputStream;
import java.io.OutputStream;
import java.net.Socket;
public class TcpClient {
    public static void main(String [] args)throws IOException {
        Socket socket = new Socket("localhost",8085);
        //获取输出流并写数据
        OutputStream outputStream = socket.getOutputStream();
        outputStream.write("get server config".getBytes());
        //获取输入流并读数据
        InputStream inputStream = socket.getInputStream();
        byte[] bytes = new byte[1024];
        int len = inputStream.read(bytes);
```

```
        String data = new String(bytes,0,len);
        System.out.println("收到服务端响应:"+data);
    }
}
```

修改主配置文件 application.yaml，加载 tcpservercfg 配置，添加如下粗体代码。

```
#Spring Boot 启动并加载 application-dev.yaml
spring:
  profiles:
    active: prefix,dev,tcpservercfg
```

先运行 Springdemo4Application，启动 Spring Boot 应用。此时会有一个线程创建 Socket 服务端并循环等待客户端请求数据，Spring Boot 应用不会自动结束。然后运行 TcpClient 的 main 方法，向服务端发送数据 get server config。服务端控制台会出现如下请求信息。

```
收到客户端数据:get server config
```

同时，TcpClient 控制台中会收到如下响应信息，信息内容与 application-tcpservercfg.yaml 配置文件的配置完全一致，证明成功获取 application-tcpservercfg.yaml 配置数据并创建 Socket 服务端。

```
收到服务端响应:TCP 服务器端口为 8085,地址是否复用为 true,接收端缓存大小为 2048
```

小　　结

本项目详细介绍了 Spring Boot 项目的创建、测试、配置、热部署、日志功能和多环境配置等内容。使读者对 Spring Boot 的使用有了初步了解。最后以一个"综合案例：用 Spring Boot 实现基于 TCP 服务的请求响应"，演示如何创建一个简单的 Spring Boot Socket 微服务和客户端进行数据交互。使读者能够体验到利用 Spring Boot 开发应用的方便和快捷。

课后练习：用 Spring Boot 实现基于 TCP 服务网购功能

模仿任务 4.6 创建一个 Spring Boot Socket 微服务项目，实现网上购物功能，商品内容通过配置文件定义。商品定义如下。

```
menu:
  saefood: yu,xia
  vegetable: qingcai,luobo
```

如海鲜包含鱼、虾等，素菜包含青菜、萝卜等。客户端模拟用户网上购物，如果客户端发送海鲜，则服务端应返回鱼、虾等内容。

项目 5 Spring Boot 数据操作和事务处理

在应用开发中,开发者通常都需要通过应用程序对数据库中数据进行读写操作。Spring 支持应用程序访问多种数据库,如关系数据库中的 MySQL、Oracle、SQLSever 和 PostgreSQL 等,以及非关系数据库中的 Redis、MongoDB 等。Spring Boot 基于 Spring 进一步简化了数据操作的编码和相关配置。本项目将以关系数据库 MySQL 为代表,介绍 Spring Boot 数据操作和事务控制相关知识,包括 Spring Boot 中主流数据库连接池的使用、Spring Boot 集成和使用 Mybatis 框架进行数据操作、事务和缓存等内容。

任务 5.1 初识数据库连接池

访问数据库的第一步就是创建数据库连接,企业级项目开发中一般会采用数据库连接池来管理所有的数据库连接。数据库连接池的作用就是为数据库连接建立一个"缓冲池"。预先在缓冲池中放入一定数量的连接,当需要操作数据时,只需从"缓冲池"中取出一个连接,使用完毕再放回去,这样避免了程序重复创建连接,增强了连接的复用性,提升了数据库性能。目前常用的数据库连接池分别是 Hikari 和 Druid,下面将分别介绍。

5.1.1 Hikari 连接池

Hikari 连接池是 Spring Boot 默认使用的连接池。使用时只需在项目的 pom.xml 文件中导入如下 spring-boot-starter-data-jdbc 依赖。

```xml
<dependency>
    <groupId>org.springframework.boot</groupId>
    <artifactId>spring-boot-starter-data-jdbc</artifactId>
</dependency>
```

Spring Boot 就同步引入了 Hikari 连接池,并进行自动配置。Hikari 连接池中的常用配置如粗体代码所示,可以自行修改默认配置,修改后的配置自动加载。

```yaml
spring:
  datasource:
    url: jdbc:mysql://localhost:3306/springdemo5?
        &useUnicode=true&characterEncoding=utf8&serverTimezone=UTC
```

```yaml
    username: root
    password: 123456
    driver-class-name: com.mysql.cj.jdbc.Driver
    #hikari 连接池配置
    hikari:
      #连接池名称,默认为 HikariPool-1
      pool-name: HikariPool-1
      #最大连接数,默认为 10
      maximum-pool-size: 15
      #连接超时时间,默认为 30 秒
      connection-timeout: 30000
      #最小空闲连接数,默认值为 10
      minimum-idle: 10
      #空闲连接最大空闲时间,默认值为 600000 毫秒
      idle-timeout: 600000
      #连接最大存活时间,一般小于数据库 wait_timeout,默认值为 1800000 毫秒
      max-lifetime: 1800000
      #测试连接是否可用,要求可查询 SQL 语句
      connection-test-query: select 1
```

如上述粗体代码修改最大连接数 maximum-pool-size 为 15,下面编码查看修改是否生效。在 Idea 中新建名为 SpringDmoe5 的 Spring Boot 项目,在创建项目时勾选 Spring data JDBC 和 MySQL Driver 依赖。项目创建完毕,pom.xml 文件依赖如下。

```xml
<dependencies>
    <dependency>
        <groupId>org.springframework.boot</groupId>
        <artifactId>spring-boot-starter-data-jdbc</artifactId>
    </dependency>
    <dependency>
        <groupId>com.mysql</groupId>
        <artifactId>mysql-connector-j</artifactId>
        <scope>runtime</scope>
    </dependency>
    <dependency>
        <groupId>org.springframework.boot</groupId>
        <artifactId>spring-boot-starter-test</artifactId>
        <scope>test</scope>
    </dependency>
</dependencies>
```

新建 Springdemo5ApplicationTests.java 测试类。在类中添加如下代码,打印 Hikari 连接池最大连接数。

```java
package springdemo5;
import com.alibaba.druid.stat.DruidStatManagerFacade;
import com.zaxxer.hikari.HikariConfig;
import org.junit.jupiter.api.Test;
```

```java
import org.springframework.beans.factory.annotation.Autowired;
import org.springframework.boot.test.context.SpringBootTest;
import org.springframework.jdbc.core.JdbcTemplate;
import java.sql.SQLException;
@SpringBootTest
class Springdemo5ApplicationTests {
    @Autowired
    private HikariConfig hikariConfig;
    @Test
    void contextLoads() {
        try{
            System.out.println(hikariConfig.getMaximumPoolSize());
        }catch(SQLException e){
            e.printStackTrace();
        }
    }
}
```

运行测试类 Springdemo5ApplicationTests，控制台打印出 15，修改的配置生效。

5.1.2 Druid 连接池

相比 Hikari 连接池，国内企业较多使用 Druid 连接池。Druid 连接池是阿里巴巴开源的一款性能高效、扩展性强的数据库连接池，该连接池结合了 C3P0、DBCP 等数据库连接池的优点，同时独特地加入了数据库监控功能，能够监控数据库连接状态和 SQL 的执行情况。

使用 Druid 连接池必须先在 pom.xml 文件中导入如下 druid-spring-boot-starter 依赖。注意这里要指定版本号，否则无法导入。版本号选择 1.2.16 版本。

```xml
<dependency>
    <groupId>com.alibaba</groupId>
    <artifactId>druid-spring-boot-starter</artifactId>
    <version>1.2.16</version>
</dependency>
```

使用 Druid 连接池只需在配置文件中配置 type 配置项，设置连接池类型为 Druid 即可，具体做法为：在 application.yaml 主配置文件中添加如下粗体代码配置 type。

```yaml
spring:
  datasource:
    url: jdbc:mysql://localhost:3306/springdemo5?&useUnicode=true
        &characterEncoding=utf8&serverTimezone=UTC
    username: root
    password: 123456
    driver-class-name: com.mysql.cj.jdbc.Driver
    #配置使用 Druid 连接池
    type: com.alibaba.druid.pool.DruidDataSource
```

然后在测试类 Springdemo5ApplicationTests 中添加如下代码，打印连接池类型。

```
System.out.println(jdbcTemplate.getDataSource().getConnection().getClass());
```

运行测试类结果如下：可以看到控制台输出连接池类型为 Druid。这时 Druid 连接池配置项使用的都是默认值。

```
class com.alibaba.druid.pool.DruidPooledConnection
```

同 Hikari 连接池一样，Druid 连接池也提供了很多配置项，如最大连接数、超时时间等。但是如果不追求性能，都可以用默认值。在测试类 Springdemo5ApplicationTests 中添加如下代码，可以打印 Druid 连接池所有配置。

```
System.out.println(DruidStatManagerFacade.getInstance().getDataSourceStatDataList());
```

在项目开发中，有时需要修改 Druid 连接池默认配置。但是 Spring Boot 本身并不会自动加载手动修改的 Druid 连接池配置。要使修改生效，需先在 application.yaml 主配置文件中进行相关配置项修改后，再手动编写 Druid 配置类加载修改后的配置。具体实现方法如下。

假设在 application.yaml 主配置文件中修改一些 Druid 连接池常用配置信息，如以下粗体代码所示。

```yaml
spring:
  datasource:
    url: jdbc:mysql://localhost:3306/springdemo5?&useUnicode=true&characterEncoding=utf8&serverTimezone=UTC&useSSL=false
    username: root
    password: 123456
    driver-class-name: com.mysql.cj.jdbc.Driver
    #配置使用Druid连接池
    type: com.alibaba.druid.pool.DruidDataSource
    #连接池名称
    name: druidpool
    #初始化连接数,默认为0
    initial-size: 10
    #最大连接数量,默认为8
    max-active: 20
    #最小连接数量
    min-idle: 2
    #获取连接等待超时的时间,单位为毫秒
    max-wait: 50000
    # 配置检测需要关闭的空闲连接的时间间隔,单位为毫秒
    time-between-eviction-runs-millis: 50000
    # 配置连接的最小生存时间
    min-evictable-idle-time-millis: 300000
    #检测连接是否有效,要求可以查询SQL语句
    validationQuery: SELECT 1 FROM DUAL
```

修改完毕，在 springdemo5 目录下新建配置文件夹 config，config 文件夹内新建 DruidConfig.java 配置类。DruidConfig 类内添加如下代码。

【DruidConfig.java】

```java
package springdemo5.config;
import com.alibaba.druid.pool.DruidDataSource;
import org.springframework.boot.context.properties.ConfigurationProperties;
import org.springframework.context.annotation.Bean;
import org.springframework.context.annotation.Configuration;
import org.springframework.stereotype.Component;
import javax.sql.DataSource;
@Configuration
public class DruidConfig {
  @Bean
  @ConfigurationProperties(prefix = "spring.datasource")
   public DataSource getDruidDataSource(){
      return new DruidDataSource();
   }
}
```

然后在测试类 Springdemo5ApplicationTests 中添加如下代码，打印 Druid 连接池所有配置。

```
System.out.println(DruidStatManagerFacade.getInstance().getDataSourceStatDataList());
```

运行测试类，在控制台可以看到如下配置信息，其中 InitialSize、MinIdle 和 MaxActive 等配置都是修改后的值，证明修改的配置已生效。

```
com. alibaba. druid. wall. WallFilter, com. alibaba. druid. filter. logging.
Slf4jLogFilter],
WaitThreadCount=0, NotEmptyWaitCount=0, NotEmptyWaitMillis=0, PoolingCount=8,
PoolingPeak=10, PoolingPeakTime=Thu Mar 16 10:24:40 CST 2023, ActiveCount=2,
ActivePeak= 2, ActivePeakTime = Thu Mar 16 10:24:40 CST 2023, InitialSize = 10,
MinIdle=2,
MaxActive=20,...
```

任务 5.2 Spring Boot Mybatis 数据操作

在本项目的任务 1 中已经介绍了 Spring Boot 使用原生 JDBC 操作数据，Spring Boot 能够极大地简化 JDBC 数据的操作。但是在企业级项目开发中通常会使用一些持久层的框架来操作数据，进一步优化数据操作性能，简化编码。Mybatis 是企业级应用开发中一款基于 Java 的持久层框架。因其简单易上手，使用灵活，已成为大多数企业构建持久层框架的首选。本任务中将重点介绍如何在 Spring Boot 项目中使用 Mybatis 持久层框架操作数据。

5.2.1 Mybatis 简介

Mybatis 是一款性能优秀的持久层框架，它提供对 JDBC 的进一步封装，通过构建 Java

实体类和数据库表之间的映射关系,将对 Java 实体类的操作转换成底层对数据库的 SQL 操作,简化了数据操作过程。Mybatis 支持动态 SQL、存储过程以及多表关联映射等功能。相比于其他持久层框架,Mybatis 有以下优势。

(1) Mybatis 配置相对简单,学习周期更短,更适合新人上手。

(2) Mybatis 是一个半自动的映射框架,提供动态 SQL,方便了 SQL 语句的编写,使数据操作更加灵活。

(3) Mybatis 提供映射标签,简化 Java 实体类与数据库字段关系映射,同时支持多表之间的关联映射,方便多表查询。

(4) Mybatis 使用 Mapper 接口类存放 Dao 层接口代码,使用 XML 配置文件或注解存放数据访问 SQL,实现 Java 代码和数据访问 SQL 语句分离,降低了 SQL 语句与程序代码的耦合,提高了代码的可维护性。

Mybatis 框架执行数据操作流程可大概分为以下四个步骤。

1. 加载配置创建会话工厂和 MappedStatement 对象

Mybatis 启动会加载 mybatis-config.xml 配置文件创建会话工厂(SqlSessionFactory),同时加载实体关系映射配置创建 MappedStatement 对象,将二者存储在内存中。其中,MappedStatement 中包括了输入参数的映射、执行的 SQL 语句、输出结果的映射等内容。

2. 组装 SQL

当收到数据操作请求时,Mybatis 会根据输入 SQL 的 ID 找到对应的 MappedStatement 对象获取 SQL 语句。将 SQL 语句和输入参数(Map、JavaBean 或者简单数据类型)进行组装,得到最终执行的 SQL 语句。

3. 执行 SQL

通过会话工厂获取一条会话(SqlSession),并通过会话获取数据库连接。在数据库连接中执行 SQL 语句操作数据并得到相应结果。Mybatis 操作数据有三种模式:SIMPLE、REUSE 和 BATCH。默认模式为 SIMPLE,即 SQL 语句在执行时每次创建一个新的预处理语句。REUSE 模式在 SQL 语句执行时复用缓存中已存在的预处理语句,提升了执行效率。BATCH 模式用于大数据量的批量更新操作,会在一次会话中批量执行所有 SQL 语句,且不需反复编译同一 SQL 语句。

4. 结果映射

将数据操作结果按照输出结果映射关系转换成特定的类型输出,类型包括 HashMap、JavaBean 和 String、Integer 等简单数据类型等。

5.2.2 Spring Boot 引入 Mybatis

要在 Spring Boot 中引入 Mybatis,需在项目 pom.xml 文件中引入如下依赖。这里指定版本为 3.0.2 版本。

```
<dependency>
    <groupId>org.mybatis.spring.boot</groupId>
    <artifactId>mybatis-spring-boot-starter</artifactId>
    <version>3.0.2</version>
</dependency>
```

引入依赖后,Spring Boot 会对 Mybatis 进行自动配置。使用时开发人员无须进行配置,只需编写 Java 实体类、Mapper 接口方法和对应的 SQL 语句即可,实现零配置使用。如需修改默认配置,在 appplication.yaml 主配置文件修改即可。在 Spring Boot 中使用 Mybatis 有两种方式:一种是基于注解,另一种是基于配置文件。这里将基于注解介绍 Mybatis 的使用。

5.2.3　Spring Boot 引入 Lombok 插件

在进行数据操作过程中,一般会通过 Java 实体类封装数据。Java 实体类内部会定义属性并根据需要提供构造方法,以及 get/set、toString、equals、hashcode 等方法。如果实体类属性过多,实体类内部代码变得冗长,不易阅读。这时就需要引入 Lombok 插件,Lombok 使用注解方式简化了上述代码的编写,将大量重复和没有技术意义的代码省略,使 Java 实体类代码整洁美观。

在 Idea 中使用 Lombok,必须要先安装 Lombok 插件,否则编辑器无法识别 Lombok 的注解。直接在工具中搜索 Lombok 插件并进行安装,如图 5-1 所示。安装完成后重启 Idea。

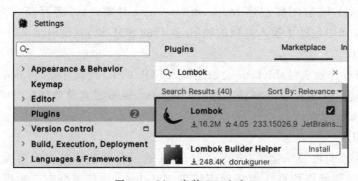

图 5-1　Idea 安装 Lombok

然后在项目的 pom.xml 文件中输入如下代码,引入 Lombok 即可使用。

```
<dependency>
    <groupId>org.projectlombok</groupId>
    <artifactId>lombok</artifactId>
</dependency>
```

Lombok 中提供了很多注解。实际开发中,一般在 Java 实体类中组合使用以下 3 个注解即可。其中,@Data 注解替代了 get/set、toString、equals、hashcode 等方法。@AllArgsConstructor 注解用于为实体类自动生成全参数的构造方法,@NoArgsConstructor 注解用于为实体类自动生成无参构造方法。

```
@Data
@AllArgsConstructor
@NoArgsConstructor
public class Demo{
    //定义一些内部属性
}
```

后续章节的相关 Java 实体类也将统一使用 Lombok 注解修饰。

5.2.4 Mybatis 注解进行单表数据操作

Mybatis 提供了 @Insert、@Update、@Delete、@Select 这 4 个注解来实现单表的 CRUD。其中，@Insert 用于执行数据新增操作，@Update 用于执行数据修改操作，@Delete 用于执行数据删除操作，@Select 用于执行数据查询操作。在 Mybatis 中利用注解操作数据，只需编写一个实体类和一个 Mapper 接口。实体类用于封装数据，Mapper 接口使用 @Mapper 和 @Repository 修饰，内部定义数据操作方法并利用注解 @Insert、@Update、@Delete、@Select 添加相应的 SQL 语句。其中，@Insert 还可以配合 @Options 注解使用返回插入数据的主键信息。数据操作方法传入参数可使用实体类对象，还可以使用 @Param 注解自由组合多个不同类型的参数，具体使用示例如下。

```
@Mapper
@Repository
public interface Mapper {
    @Insert("插入语句")
    @Options(useGeneratedKeys = true,keyProperty = "主键属性")
    Insert 方法名(实体类 实体类对象);
    @Update("修改语句")
    Update 方法名(实体类 实体类对象);
    @Select("查询语句")
    Select 方法名(实体类 实体类对象);
    @Delete("删除语句")
    Delete 方法名(实体类 实体类对象);
    @Select("查询语句")
    Select 方法名(@Param ("参数别名1") 数据类型1 参数名1,
                  @Param ("参数别名2") 数据类型2 参数名2);
}
```

如果实体类对象属性名和数据库表字段名不一致，可以组合使用 @Result、@Results、@ResultMap 注解配置属性名和表字段的映射关系，示例如下。

```
@Insert("插入语句")
@Results(id="resMap",value={
    @Result(column="字段1", property="属性1", jdbcType= 数据类型, 字段1=true),
    @Result(column="字段2", property="属性2", jdbcType=数据类型),
    ...
})
Insert 方法名(实体类 实体类对象);
@Insert("插入语句")
@ResultMap(value="resMap")        //引用 id 为 resMap 的映射关系
Update 方法名(实体类 实体类对象);
```

其中，@Results 的 id 属性表示映射的唯一标识，value 属性表示实体类对象属性名和数据库表字段名之间具体的映射关系。@Result 注解用于表示一个属性名和一个表字段名之间的映射关系,内部 column 属性为数据库字段名,property 属性为实体类属性名,jdbcType 属性为数据库字段类型。true 表示该字段为主键,默认为 false。@ResultMap 用于引用已定义的映射关系,value 属性值为已有映射的 id。@ResultMap 可减少代码量,提高代码的复用性。但是在企业级项目开发中为了方便,建议还是将实体类对象属性名和数据库表字段名保持一致。

下面演示@Insert、@Update、@Delete、@Select 注解的具体用法。

假设已安装好 MySQL 8。在 MySQL 8 中创建名为 springdemo5 的数据库,数据库字符集编码设为 utf8。执行如下 SQL 语句,创建一个图书信息表 book_info,其中包含 id、book_id、book_name、book_price 和 book_author 五列。

```sql
CREATE TABLE IF NOT EXISTS book_info (
    id    int(11) NOT NULL AUTO_INCREMENT,
    book_id   int(10) NOT NULL,
    book_name   varchar(50) NOT NULL,
    book_price  decimal(6,1) NOT NULL,
    book_author   varchar(50) NOT NULL,
    PRIMARY KEY (id)
);
```

执行如下 SQL 语句,先向 book_info 表中添加两条数据,如图 5-2 所示。

```sql
insert into book_info (book_id,book_name,book_price,book_author)
values(1,"book1",20.5,"author1");
insert into book_info (book_id,book_name,book_price,book_author)
values(2,"book2",30.5,"author2");
```

id	book_id	book_name	book_price	book_author
1	1	book1	20.5	author1
2	2	book2	30.5	author2

图 5-2 插入两条数据

在 springdemo5 目录下新建 domain 文件夹,在 domain 文件夹下新建 Book 类并使用 Lombok 注解修饰,用于封装查询数据。为方便 Book 类属性名和表字段名保持完全一致。如果属性名和表字段名不一致,可在属性上使用@Column(value = "表字段名")注解映射。

【Book.java】

```java
package springdemo5.domain;
import lombok.AllArgsConstructor;
import lombok.Data;
import lombok.NoArgsConstructor;
@Data
@AllArgsConstructor
@NoArgsConstructor
```

```
public class Book {
    private Integer id;
    private Integer book_id;
    private String book_name;
    private Double book_price;
    private String book_author;
}
```

在 springdemo5 目录下新建 mapper 文件夹，在 mapper 文件夹下新建 BookMapper 接口，内部定义增、删、改、查方法并添加对应的 SQL 语句。其中，insert 方法中使用@Result 演示配置实体类对象属性名和数据库表字段名的映射关系，并使用@Options(useGeneratedKeys = true,keyProperty = "id")返回插入数据的主键信息。update 方法使用@ResultMap 引用 insert 方法中配置的映射关系。

【BookMapper.java】

```
package springdemo5.mapper;
import org.apache.ibatis.annotations.*;
import org.apache.ibatis.type.JdbcType;
import org.springframework.stereotype.Repository;
import springdemo5.domain.Book;
import java.util.List;
@Mapper
@Repository
public interface BookMapper {
    //新增数据
    @Insert("insert into book_info (book_id,book_name,book_price,book_author)" +
            "values(#{book_id},#{book_name},#{book_price},#{book_author})")
    //定义属性名和表字段映射关系,映射关系 id 为 resMap
    @Results(id="resMap",value={
        @Result(column="id", property="id", jdbcType= JdbcType.INTEGER, id=true),
        @Result(column="book_id", property="book_id",
                jdbcType=JdbcType.INTEGER),
        @Result(column="book_name ", property="book_name",
                jdbcType=JdbcType.VARCHAR),
        @Result(column="book_price ", property="book_price",
                jdbcType=JdbcType.DECIMAL),
        @Result(column="book_author ", property="book_author",
                jdbcType=JdbcType.VARCHAR)
    })
    @Options(useGeneratedKeys = true,keyProperty = "id")   //返回主键
    int insertBook(Book book);
    //修改数据
    @Update("update book_info set book_name=#{book_name} " +
            "where book_id=#{book_id}")
    @ResultMap(value = "resMap")//引用 id 为 resMap 的映射
    int updateBook(Book book);
    //模糊查询
    @Select("select * from book_info where book_name " +
```

```
            "like concat('%',#{book_name},'%')")
    List<Book> selectBookByName(Book book);
    //自由组合输入参数查询数据
    @Select("select * from book_info where book_name " +
            "=#{book_name} and book_author=#{book_author}")
    List<Book> selectBookByNameAndAuthor(@Param("book_name")String name,
            @Param("book_author") String author);
    //删除数据
    @Delete("delete from book_info where book_name=#{book_name}")
    int deleteBook(Book book);
}
```

利用 4.3.2 小节的方法创建 BookMapper 的测试类 BookMapperTest.java,在测试类中编写 insertBook 方法,测试新增功能。

【BookMapperTest.java】

```
package springdemo5.mapper;
import org.junit.jupiter.api.Test;
import org.springframework.beans.factory.annotation.Autowired;
import org.springframework.boot.test.context.SpringBootTest;
import springdemo5.domain.Book;
@SpringBootTest
class BookMapperTest {
    @Autowired
    private BookMapper bookMapper;
    @Test
    void insertBook() {
        Book book=new Book();
        book.setBook_id(4);
        book.setBook_name("book4");
        book.setBook_price(20.5);
        book.setBook_author("author4");
        bookMapper.insertBook(book);
        System.out.println("插入数据主键为"+book.getBook_id());
    }
}
```

运行 insertBook 方法,控制台打印信息:插入数据主键为 4。同时数据库新增一条记录,结果如图 5-3 所示。

id	book_id	book_name	book_price	book_author
1	1	book1	20.5	author1
2	2	book2	30.5	author2
3	3	book3	20.5	author3

图 5-3 Mybatis 新增数据

在测试类中编写 updateBook 方法,测试修改功能。修改 book_id 为 4 的 book_name 值为 book5。

```
@Test
void updateBook() {
    Book book=new Book();
    book.setBook_id(3);
    book.setBook_name("book5");
    bookMapper.updateBook(book);
}
```

运行 updateBook 方法,数据修改成功,结果如图 5-4 所示。

id	book_id	book_name	book_price	book_author
1	1	book1	20.5	author1
2	2	book2	30.5	author2
3	3	book5	20.5	author3

图 5-4 Mybatis 修改数据

在测试类中编写 selectBookByName 方法,测试查询功能。模糊查询 book_name 值包含"book"的数据。

```
@Test
void selectBookByName() {
    Book book=new Book();
    book.setBook_name("book");
    System.out.println(bookMapper.selectBookByName(book));
}
```

运行 selectBookByName 方法,数据查询结果如下。

```
[Book(id=1, book_id=1, book_name=book1, book_price=20.5, book_author=author1),
Book(id=2, book_id=2, book_name=book2, book_price=30.5, book_author=author2),
Book(id=3, book_id=3, book_name=book5, book_price=20.5, book_author=author3)]
```

在测试类中编写 selectBookByNameAndAuthor 方法,测试查询功能。查询 book_name 为 book1 且 book_author 为 author1 的数据。

```
@Test
void selectBookByNameAndAuthor(){
    String bookName="book1";
    String bookAuthor="author1";
    System.out.println(bookMapper.selectBookByNameAndAuthor(bookName,
bookAuthor))
}
```

运行 selectBookByNameAndAuthor 方法,数据查询结果如下。

```
[Book(id=2, book_id=1, book_name=book1, book_price=20.5, book_author=author1)]
```

在测试类中编写 deleteBook 方法,测试删除功能。删除 book_name 为 book5 的数据。

```
@Test
void deleteBook() {
    Book book=new Book();
    book.setBook_name("book5");
    bookMapper.deleteBook(book);
}
```

运行 deleteBook 方法,book_name 为 book5 的数据已删除,结果如图 5-5 所示。

▼ WHERE			≡▼ ORDER BY	
id	book_id	book_name	book_price	book_author
1	1	book1	20.5	author1
2	2	book2	30.5	author2

图 5-5 Mybatis 删除数据

5.2.5 Mybatis 注解进行多表关联查询

在数据查询过程中,除了单表的数据查询外,还有多表的关联查询。在数据库中多表关联有三种关系,分别是一对一、一对多和多对多。

一对一:如人员信息表和身份证信息表就是一对一关系。一个人对应一张身份证,一张身份证对应一个人。一对一关系特征为在任意一方引入对方主键作为外键。

一对多:如班级信息表和学生信息表就是一对多关系,一个班级对应多名学生。一对多关系特征为在"多"方引入"一"方的主键作为外键。

多对多:如订单信息表和商品信息表就是多对多关系,一个订单对应多个商品。一个商品也可以对应多个订单。多对多关系一般需引入中间表,在中间表中引入双方主键作为外键,中间表的主键使用两个外键做联合主键或新的字段作为单独主键。

如果用实体类 A 和 B 来描述,一对一、一对多和多对多可以表示成如图 5-6 所示形式。

一对一为在本类中定义对方类型的属性。可在 A 类中定义 B 类类型的属性 b,或在 B 类中定义 A 类类型的属性 a。使用时主要看业务需求,业务需求查询发起者是 A 类,则先查 A 后查 B,在 A 类中定义 B 类型属性,反之亦然。

一对多为"一"方类中定义"多"方类型的集合属性,

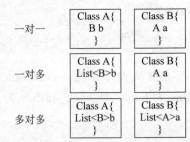

图 5-6 实体类描述关联模式

也可在"多"方类中定义"一"方属性,使用时主要看业务需求,如果业务需求查询发起者是"一"方,在"一"方类中定义"多"方类型的集合属性。如果业务需求查询发起者是"多"方,在"多"方类中定义"一"方属性。

多对多为 A 和 B 类对象互为对方的集合属性。使用时主要看业务需求,业务需求查询发起者是 A 类,则先查 A 后查 B,在 A 类中定义 B 类型集合属性,反之亦然。

编程中,实现关联映射只需在当前实体类添加相应的关联属性,同时在与之对应的 Mapper 类查询方法上使用如下@Result 注解对关联属性赋值。

数据库中多表关联三种关系

```
@Result(column ="",property = "",javaType = "",
        one = @One(select="")或many = @Many(select=""))
```

其中,column 表示用当前类对应的数据表中的哪个字段去执行关联查询,如果关联查询需要传多个字段值,可采用键值对写法传值,具体示例如下。

```
column = "{userName=user_name,userId=user_id}"
```

userName 和 userId 为键,user_name 和 user_id 为字段名,同时关联表中以 Map 形式接收参数,使用♯{userName}取 user_name 字段值,♯{userId}取 user_id 字段值。property 表示当前类中的关联属性,javaType 表示查询数据返回类型,一般为某实体类或实体类集合。one=@One 或 many=@Many(select="")中的 select 用于指定调用哪个类的方法执行关联查询。下面介绍一对一、一对多和多对多三种模式的编程实现。

1. 一对一

一对一映射通过在@Result 注解中设置 one=@One()实现。这里以人员信息表和身份证信息表为例实现一对一关系,一个人员对应一个身份证。查询时从人员信息表发起查询,查询所有人员信息,然后利用人员 id 去查询身份证信息。先在数据库 springdemo5 中利用如下 SQL 语句创建 Person 表和 Card 表并添加数据,Card 表中引入外键 c_personid。

```sql
# 人员信息表
CREATE TABLE IF NOT EXISTS person(
    p_id INT NOT NULL auto_increment,
    p_name VARCHAR(50) NOT NULL,
    PRIMARY KEY(p_id)
);
INSERT INTO person (p_name) VALUES('person1');
INSERT INTO person (p_name) VALUES('person2');
# 身份证信息表
CREATE TABLE IF NOT EXISTS card(
    c_id INT NOT NULL auto_increment,
    c_code VARCHAR(50) NOT NULL,
    c_createtime TIMESTAMP NOT NULL,
    c_personid INT NOT NULL,
    PRIMARY KEY(c_id),
    FOREIGN KEY(c_personid) REFERENCES person(p_id)
);
INSERT INTO card (c_code,c_personid,c_createtime)
VALUES('12345',1,'2023-03-19 12:00:00');
INSERT INTO card (c_code,c_personid,c_createtime)
VALUES('56789',2,'2023-03-20 12:00:00');
```

在 domain 目录下新建实体类 Person 和 Card，Person 类内部定义常规属性外还添加了 card 属性，用于一对一映射。

【Person.java】

```java
package springdemo5.domain;
import lombok.AllArgsConstructor;
import lombok.Data;
import lombok.NoArgsConstructor;
import java.util.List;
@Data
@AllArgsConstructor
@NoArgsConstructor
public class Person {
    private Integer p_id;
    private String p_name;
    //一对一属性
    private Card card;
}
```

【Card.java】

```java
package springdemo5.domain;
import lombok.AllArgsConstructor;
import lombok.Data;
import lombok.NoArgsConstructor;
@Data
@NoArgsConstructor
@AllArgsConstructor
public class Card {
    private Integer c_id;
    private String c_code;
    private String c_createtime;
}
```

在 mapper 文件夹新建 PersonMapper 和 CardMapper 接口类，类上分别添加@Mapper 和@Repository 注解。PersonMapper 内部定义方法 selectall 查询所有人员信息，并使用@One 注解和身份证进行一对一映射。

【PersonMapper.java】

```java
package springdemo5.mapper;
import org.apache.ibatis.annotations.*;
import org.springframework.stereotype.Repository;
import springdemo5.domain.Card;
import springdemo5.domain.Person;
import java.util.List;
@Mapper
@Repository
public interface PersonMapper {
```

```
    @Select("select * from person")
    @Results({
            @Result(column = "p_id",property = "p_id"),
            @Result(column = "p_id",property = "card",javaType =Card.class,
                    one=@One(select="springdemo5.mapper.CardMapper.selectbyId"))
    })
    List<Person> selectAll();
}
```

CardMapper 中定义 selectbyId 方法,通过人员 id 查询身份证信息。

【CardMapper.java】

```
package springdemo5.mapper;
import org.apache.ibatis.annotations.*;
import org.springframework.stereotype.Repository;
import springdemo5.domain.Card;
@Mapper
@Repository
public interface CardMapper {
    //根据输入的 p_id 查询 Card 对象数据
    @Select("select * from card where c_personid=#{c_personid}")
    Card selectbyId(Integer c_personid);
}
```

以上粗体代码实现了一对一映射,首先利用 select * from person 查询所有人员信息,并赋值给 Person 对象,然后将 p_id 值赋值给 Person 对象的 p_id 属性,并利用 p_id 值输入 CardMapper 的 selectbyId 方法查询身份证信息并赋值给 card 属性。

利用 PersonMapper 生成测试类 PersonMapperTest,在测试类中调用 selectAll 方法查询数据。

【PersonMapperTest.java】

```
package springdemo5.mapper;
import org.junit.jupiter.api.Test;
import org.springframework.beans.factory.annotation.Autowired;
import org.springframework.boot.test.context.SpringBootTest;
@SpringBootTest
class PersonMapperTest {
    @Autowired
    private PersonMapper personMapper;
    @Test
    void selectAll() {
        System.out.println(personMapper.selectAll());
    }
}
```

运行测试类,控制台输出的如下人员信息和身份证信息是一对一映射数据。

```
[Person(p_id=1, p_name=person1, card=Card(c_id=1, c_code=12345,
c_createtime=2023-03-19 12:00:00)), Person(p_id=2, p_name=person2, card=Card
(c_id=2,
c_code=56789, c_createtime=2023-03-20 12:00:00))]
```

2. 一对多

一对多映射通过在@Result注解中设置many = @Many()实现。这里以人员信息和订单信息实现一对多关系,一个人对应多个订单。查询时从人员信息表发起查询,查询所有人员信息,然后利用人员id去查询订单信息。使用如下SQL语句在springdemo5数据库中创建订单表并添加数据。

```sql
# 订单信息表
CREATE TABLE IF NOT EXISTS orders(
    o_id INT NOT NULL auto_increment,
    o_name VARCHAR(50) NOT NULL,
    o_personid INT NOT NULL,
    PRIMARY KEY(o_id),
    FOREIGN KEY(o_personid) REFERENCES person(p_id)
);
INSERT INTO orders (o_name,o_personid) VALUES('order1',1);
INSERT INTO orders (o_name,o_personid) VALUES('order2',1);
INSERT INTO orders (o_name,o_personid) VALUES('order3',2);
INSERT INTO orders (o_name,o_personid) VALUES('order4',2);
```

修改Person类,内部新增的属性orders用于一对多映射。

```java
private List<Orders> orders;
```

在domain文件夹下新建实体类Orders。

【Orders.java】

```java
package springdemo5.domain;
import lombok.AllArgsConstructor;
import lombok.Data;
import lombok.NoArgsConstructor;
import java.util.List;
@Data
@AllArgsConstructor
@NoArgsConstructor
public class Orders {
    private Integer o_id;
    private String o_name;
}
```

修改PersonMapper类,在selectAll方法的@Results注解内部新增如下代码:

```java
@Result(column ="p_id",property = "orders",javaType = List.class,
        many = @Many(select="springdemo5.mapper.OrderMapper.selectOrderById"))
```

以上代码用于在一对多关系中利用 p_id 值输入 OrderMapper 的 selectOrderById 方法查询数据,并将结果赋值给 orders 属性。

在 mapper 文件夹下新建 OrderMapper 类,内部定义的 selectOrderById 方法接收 p_id 值,查询该人员的订单信息。

【OrderMapper.java】

```java
package springdemo5.mapper;
import org.apache.ibatis.annotations.*;
import org.springframework.stereotype.Repository;
import springdemo5.domain.Orders;
import java.util.List;
@Mapper
@Repository
public interface OrderMapper {
    @Select("select * from orders where o_personid=#{o_personid}")
    List<Orders> selectOrderById(Integer o_personid);
}
```

运行测试类 PersonMapperTest,控制台输出的如下人员信息和身份证信息是一对一映射数据,人员信息和订单信息是一对多映射数据。

```
[Person(p_id=1, p_name=person1, card=Card(c_id=1, c_code=12345,
c_createtime=2023-03-19 12:00:00), orders=[Orders(o_id=1, o_name=order1),
Orders(o_id=2, o_name=order2)]),
Person(p_id=2, p_name=person2, card=Card(c_id=2, c_code=56789,
c_createtime=2023-03-20 12:00:00), orders=[Orders(o_id=3, o_name=order3),
Orders(o_id=4, o_name=order4)])]
```

3. 多对多

多对多映射需创建中间表,实现时需转为一对多映射,以一个"多"方为基础使用关联查询来查询中间表数据,并在另外方的@Result 注解中设置 many = @Many()实现。这里将订单信息和商品信息实现多对多关系,多个订单对应多个商品。查询时从订单信息表发起查询,查询所有订单信息,然后利用订单 id 去查询商品信息。使用如下 SQL 语句在 springdemo5 数据库中创建商品信息表和订单商品表并添加数据。

```sql
# 商品信息表
CREATE TABLE IF NOT EXISTS product(
    pro_id INT NOT NULL auto_increment,
    pro_name VARCHAR(50) NOT NULL,
    PRIMARY KEY(pro_id)
);
INSERT INTO product (pro_name) VALUES('product1');
INSERT INTO product (pro_name) VALUES('product2');
# 订单商品表
CREATE TABLE IF NOT EXISTS order_product(
    ord_pro_id INT NOT NULL auto_increment,
    ord_pro_oid INT NOT NULL,
```

```
    ord_pro_pid INT NOT NULL,
    PRIMARY KEY(ord_pro_id)
);
INSERT INTO order_product (ord_pro_oid,ord_pro_pid) VALUES(1,1);
INSERT INTO order_product (ord_pro_oid,ord_pro_pid) VALUES(1,2);
INSERT INTO order_product (ord_pro_oid,ord_pro_pid) VALUES(2,1);
INSERT INTO order_product (ord_pro_oid,ord_pro_pid) VALUES(2,2);
INSERT INTO order_product (ord_pro_oid,ord_pro_pid) VALUES(3,1);
INSERT INTO order_product (ord_pro_oid,ord_pro_pid) VALUES(3,2);
INSERT INTO order_product (ord_pro_oid,ord_pro_pid) VALUES(4,1);
INSERT INTO order_product (ord_pro_oid,ord_pro_pid) VALUES(4,2);
```

修改 Orders 类，内部新增的属性 productList 用于多对多映射。

```
List<Product> productList;
```

在 domain 文件夹下新建实体类 Product。

【Product.java】

```
package springdemo5.domain;
import lombok.AllArgsConstructor;
import lombok.Data;
import lombok.NoArgsConstructor;
@Data
@AllArgsConstructor
@NoArgsConstructor
public class Product {
    private Integer pro_id;
    private String pro_name;
}
```

修改 OrderMapper，在 selectOrderById 方法上添加如下 @Results 注解。

```
@Results({
        @Result(property = "o_id",column = "o_id"),
        @Result(property = "productList",column = "o_id",javaType = List.class,
    many=@Many(select = "springdemo5.mapper.ProductMapper.selectProductByOid"))
})
```

以上代码用于多对多关系中，可利用 o_id 值输入 ProductMapper 的 selectProductByOid 方法查询数据，并将结果赋值给 productList 属性。

在 mapper 文件夹下新建 ProductMapper，内部定义的 selectProductByOid 方法接收 o_id 值并查询该订单的商品信息。

【ProductMapper.java】

```
package springdemo5.mapper;
import org.apache.ibatis.annotations.Mapper;
```

```java
import org.apache.ibatis.annotations.Select;
import org.springframework.stereotype.Repository;
import springdemo5.domain.Product;
import java.util.List;
@Mapper
@Repository
public interface ProductMapper {
    @Select("select * from product inner join order_product on " +
            "product.pro_id=order_product.ord_pro_pid " +
            "where order_product.ord_pro_oid=#{ord_pro_oid}")
    List<Product> selectProductByOid(Integer ord_pro_oid);
}
```

运行测试类 PersonMapperTest,控制台输出的如下人员信息和身份证信息是一对一映射数据,人员信息和订单信息是一对多映射数据,订单信息和商品信息是多对多映射数据。

```
[Person(p_id=1, p_name=person1, card=Card(c_id=1, c_code=12345,
c_createtime=2023-03-19 12:00:00), orders=[Orders(o_id=1, o_name=order1,
productList=[Product(pro_id=1, pro_name=product1),
Product(pro_id=2, pro_name=product2)]), Orders(o_id=2, o_name=order2,
productList=[Product(pro_id=1, pro_name=product1), Product(pro_id=2,
pro_name=product2)])]),
Person(p_id=2, p_name=person2, card=Card(c_id=2, c_code=56789,
c_createtime=2023-03-20 12:00:00), orders=[Orders(o_id=3, o_name=order3,
productList=[Product(pro_id=1, pro_name=product1),
Product(pro_id=2, pro_name=product2)]), Orders(o_id=4, o_name=order4,
productList=[Product(pro_id=1, pro_name=product1), Product(pro_id=2,
pro_name=product2)])])]
```

5.2.6 Mybatis 注解动态 SQL

在数据操作过程中,经常会遇到这样的情况:以数据查询为例,需要通过随机组合多个查询条件进行数据查询。因为并不知道用户会具体使用哪些查询条件,往往需要采用枚举法写大量的 SQL 语句来满足所有的查询组合。如果由于业务需求又新增若干查询条件,那么最终 SQL 语句的数量将呈现线性增长。这种现象也同样发生在数据新增、修改和删除的场景中。为了解决此类问题,Mybatis 提供了动态 SQL 功能。该功能基于 OGNL 表达式允许开发人员根据实际需求灵活拼接 SQL 语句,避免了数据操作过程中单一 SQL 语句的重复使用,提高了 SQL 语句的复用性。

Mybatis 基于注解实现动态 SQL 有两种方式:一种是基于 XML 动态标签,另一种是基于 @*Provider 注解。下面分别介绍二者的具体实现。

1. 基于 XML 动态标签

Mybatis 提供一些常用的 XML 动态标签实现动态 SQL,这些标签如表 5-1 所示。

表 5-1　Mybatis 动态 SQL 常用标签

标签	描述
\<if\>	类似 if 条件判断语句，用于单条件判断
\<choose\>(\<when\>,\<otherwise\>)	类似 Java 的 switch case 语句，用于多条件判断
\<where\>	用于 SQL 语句条件的动态拼接
\<trim\>	用于在动态拼接 SQL 语句中去除特定的字符串
\<set\>	用于动态拼接 update 语句的字段
\<foreach\>	用于循环遍历集合数据取值

下面介绍表 5-1 所示标签的使用方法。

（1）\<if\>标签。该标签用于在数据操作中对字段值进行单个条件判断，如果满足条件则将该字段动态拼接进 SQL 语句，否则不拼接。\<if\>标签最常用的场景就是数据操作时判断字段值是否为空或是否合法。这里以查询场景演示\<if\>标签的使用。

在 ProductMapper 类中添加如下 selectProductByIdAndName 方法，使用\<if\>标签组合 pro_id 和 pro_name 查询产品信息。pro_id 有值就拼接 pro_id，pro_name 有值就拼接 pro_name，二者都有值就同时拼接。

```
@Select("<script>" +
    "select * from product where 1=1 " +
    "<if test = 'pro_id != null'>" +
    "and pro_id = #{pro_id}" +
    "</if>" +
    "<if test = 'pro_name != null'>" +
    "and pro_name = #{pro_name}" +
    "</if>" +
    "</script>")
Product selectProductByIdAndName(Product product);
```

其中，整个 SQL 语句必须包含在\<script\>\</script\>标签中，where 1=1 是为了防止当 pro_id 和 pro_name 都不给值的情况下，SQL 语句为不会组装成 select * from product where 而报错。

为了能在控制台看到动态组装的 SQL 语句，这里在 application.yaml 中添加如下配置，设置 springdemo5 的 mapper 目录下所有文件执行的日志为 DEBUG 级别。

```
logging:
  level:
    springdemo5.mapper: DEBUG
```

创建测试类 ProductMapperTest，在内部输入以下代码，调用 selectProductByIdAndName 方法。

```
package springdemo5.mapper;
import org.junit.jupiter.api.Test;
import org.springframework.beans.factory.annotation.Autowired;
```

```
import org.springframework.boot.test.context.SpringBootTest;
import springdemo5.domain.Product;
@SpringBootTest
class ProductMapperTest {
    @Autowired
    private ProductMapper productMapper;
    @Test
    void selectProductByIdAndName() {
        //通过 pro_id 和 pro_name 同时查询
        productMapper.selectProductByIdAndName(new Product(1,"product1"));
        //只通过 pro_name 查询
        Product product=new Product();
        product.setPro_name("product1");
        productMapper.selectProductByIdAndName(product);
    }
}
```

运行测试类,在控制台打印如下 SQL 信息,证明查询时<if>标签按照字段是否为空进行了动态拼接。

```
Preparing: select * from product where 1=1 and pro_id = ? and pro_name = ?
Parameters: 1(Integer), product1(String)
Preparing: select * from product where 1=1 and pro_name = ?
Parameters: product1(String)
```

(2)<choose>(<when>,<otherwise>)标签。在数据操作过程中,如果字段之间存在优先级关系。按字段优先级从高到低排序,如果优先级最高的字段满足条件被拼接,则其他优先级低的所有字段不再拼接。如果优先级最高的字段不满足条件没有被拼接,再看优先级次高字段,并以此类推。这种情况利用<if>标签是无法实现的,因为<if>标签的所有字段都是等价的,只要条件为真就可以被拼接。这时就可以利用<choose>(<when>,<otherwise>)标签,<choose>(<when>,<otherwise>)执行流程类似于 Java 中的 swith case default 语句,只要有一个 when 满足条件可以拼接字段,后续 when 条件不再判断;如果所有 when 条件都不满足,则会拼接 otherwise 里面的语句。

这里在 ProductMapper 类中添加如下 selectProductByChoose 方法,使用<choose>(<when>,<otherwise>)标签查询产品信息,优先拼接 pro_id 字段,若 pro_id 字段为空,则拼接 pro_name。selectProductByChoose 方法返回值类型为 List<Product>。

```
@Select("<script>" +
    "select * from product where 1=1 " +
    "<choose>" +
    "<when test = 'pro_id != null'>" +
    "and pro_id = #{pro_id}" +
    "</when>" +
    "<when test = 'pro_name != null'>" +
    "and pro_name = #{pro_name}" +
    "</when>" +
```

```
    "<otherwise>" +
    "</otherwise>" +
"</choose>" +
"</script>")
List<Product> selectProductByChoose(Product product);
```

在测试类 ProductMapperTest 中输入以下代码,调用 selectProductByChoose 方法。

```
//pro_id 和 pro_name 同时有值
productMapper.selectProductByChoose(new Product(1,"product1"));
//pro_id 和 pro_name 都没有值
Product product=new Product();
productMapper.selectProductByChoose(product);
//pro_id 为空,pro_name 有值
product.setPro_name("product1");
productMapper.selectProductByChoose(product);
```

运行测试类,在控制台打印如下 SQL 信息,证明查询时<choose>(<when>,<otherwise>)标签按照字段优先级进行了动态拼接。如果 pro_id 和 pro_name 同时有值,优先拼接 pro_id。pro_id 为空,pro_name 有值则拼接 pro_name,如果 pro_id 和 pro_name 都没有值,则不拼接。

```
Preparing: select * from product where 1=1 and pro_id = ?
Parameters: 1(Integer)
Preparing: select * from product where 1=1
Parameters:
Preparing: select * from product where 1=1 and pro_name = ?
Parameters: product1(String)
```

(3) <where>标签。在上述两个案例中都引入了 where 1=1 的拼接条件,但是在写 SQL 语句时,我们又不习惯在 where 后添加 1=1,这样会使 SQL 语句显得很奇怪。如果去掉 1=1,动态拼接的 SQL 语句就变成如下形式。

```
select * from product where and pro_id = ? and pro_name = ?
```

这显然是会报错的。那么,如何在去掉 1=1 的情况下又将多余 and 删除呢?这就需要使用<where>标签。<where>标签用于替换 where 1=1 的拼接条件,<where>标签会自动判断内部条件是否成立,如果成立则拼接,拼接完毕可以自动去除 SQL 语句中多余的 and 或 or 关键字。如可以将 ProductMapper 类中的 selectProductByIdAndName 方法改成如下 SQL 语句,一样能够正常查询数据。

```
@Select("<script>" +
    "select * from product " +
    "<where>" +
    "<if test = 'pro_id != null'>" +
    "and pro_id = #{pro_id}" +
```

```
    "</if>" +
    "<if test = 'pro_name != null'>" +
    "and pro_name = #{pro_name}" +
    "</if>" +
    "</where>" +
    "</script>")
Product selectProductByIdAndName(Product product);
```

(4)〈trim〉。〈where〉能够自动去除 SQL 语句中多余的 and 或 or 关键字。那么如果在 SQL 语句中还有一些多余的其他特殊字符串需要删除,如何解决?此时就需要用到〈trim〉标签。〈trim〉标签提供了 prefix、suffix、prefixOverrides 和 suffixOverrides 等常用属性,用于去除 SQL 内部某段范围内指定的字符串。其中 prefix 和 suffix 一般配合使用,指定处理的 SQL 语句范围起始和结束位置字符。prefixOverrides 用于指定 SQL 语句要去除的第一个字符,可结合 prefix 和 suffix 使用,并指定 SQL 语句处理范围。suffixOverrides 用于指定 SQL 语句要去除的最后一个字符,可结合 prefix 和 suffix 使用,并指定 SQL 语句处理范围。

下面演示〈trim〉标签的具体用法,这里使用〈trim〉标签写一条插入语句,对产品信息表插入一条记录。

在 ProductMapper 类中添加如下 insertProduct 方法,新增产品记录。

```
@Insert("<script>" +
    "insert into product" +
    "<trim prefix = '(' suffix = ')' suffixOverrides=','>" +
    "<if test = 'pro_id != null'> " +
    "pro_id," +
    "</if>" +
    "<if test = 'pro_name != null'> " +
    "pro_name," +
    "</if>" +
    "</trim>" +
    "<trim prefix = 'values (' suffix = ')' suffixOverrides=','>" +
    "<if test = 'pro_id != null'> " +
    "#{pro_id}," +
    "</if>" +
    "<if test = 'pro_name != null'> " +
    "#{pro_name}," +
    "</if>" +
    "</trim>" +
    "</script>" )
@Options(useGeneratedKeys = true,keyProperty = "pro_id")//返回主键
int insertProduct(Product product);
```

这里使用了两次〈trim〉标签,第一次〈trim〉标签使用 prefix、suffix 属性组合指定 SQL 语句范围为 insert into product 后面一对小括号内部,并使用 suffixOverrides 属性对小括号内部 SQL 语句中最后一个逗号进行去除。第二次〈trim〉标签使用 prefix、suffix 属性组合指定 SQL 语句范围为 values 的小括号内部,并使用 suffixOverrides 属性对小括

号内部 SQL 语句中最后一个逗号进行去除。如果实际开发中需要对指定范围内 SQL 语句第一个字符串进行去除,可以使用 prefixOverrides。

在测试类 ProductMapperTest 中添加如下代码,调用 insertProduct 方法。

```
Product newProduct=new Product();
newproduct.setPro_name("product3");
productMapper.insertProduct(newproduct);
System.out.println("插入数据主键为:"+newProduct.getPro_id());
```

运行测试类,在控制台打印如下 SQL 信息,记录插入成功。

```
Preparing: insert into product ( pro_name ) values ( ? )
Parameters: product3(String)
Updates: 1
插入数据主键为:3
```

数据库中也能看到 product3 产品信息,如图 5-7 所示。

(5) <set>。<set>标签主要用于数据更新操作。<set>标签替代了 update 语句中的 set 关键字,并根据内部字段是否满足条件进行 SQL 语句的动态拼接,同时去除末尾多余的逗号。下面演示<set>标签的使用,修改 pro_id 为 3 的产品名称 pro_name 为 product4。

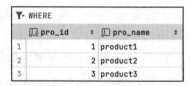

图 5-7 新增产品信息

在 ProductMapper 类中添加如下 updateProduct 方法,修改产品记录。

```
@Update("<script>" +
    "update product" +
    "<set>" +
    "<if test = 'pro_name != null'> " +
    "pro_name = #{pro_name}" +
    "</if>" +
    "</set>" +
    "<where>" +
    "<if test = 'pro_id != null'> " +
    "and pro_id = #{pro_id}" +
    "</if>" +
    "</where>" +
    "</script>")
int updateProduct(Product product);
```

在测试类 ProductMapperTest 中添加如下代码,调用 updateProduct 方法。

```
productMapper.updateProduct(new Product(3,"product4"));
```

运行测试类,在控制台打印如下 SQL 语句信息,记录修改成功。

```
Preparing: update product SET pro_name=? WHERE pro_id=?
```

```
Parameters: product4(String), 3(Integer)
Updates: 1
```

数据库中也能看到 pro_id 为 3 的产品名称 pro_name 被修改成了 product4，如图 5-8 所示。

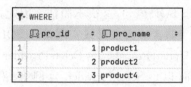

图 5-8　修改产品信息

注意：在使用＜set＞标签更新数据时，需确保内部字段至少有一个满足条件并能够被拼接，如果所有字段都不满足条件，则 SQL 语句执行失败。

（6）＜foreach＞。如果 SQL 语句中含有 in 条件，需要在集合范围内查询数据，例如查询 id 为 1~10 的数据，这种情况可以使用 foreach 来实现。foreach 语法如下。

```
<foreach item="" index="" collection="" open="" separator="" close="">
    参数值
</foreach>
```

其中，item 表示迭代时的当前元素值。index 表示每次迭代的索引。collection 表示查询的集合数据，在不同情况下，collection 值会不一样，如果方法传入的是单参数且参数类型是 List，collection 值为 list；如果方法传入的是单参数且参数类型是 Array 数组，collection 值为 array；如果方法传入的参数是多个，就需要把它们封装成一个 Map。collection 值是 Map 的键。open 表示该语句以什么开始，如果是 in 条件语句，以"("开始。separator 表示在每次迭代时以什么符号作为分隔符，如果是 in 条件语句，以","作为分隔符。close 表示该语句以什么结束，如果是 in 条件语句，以")"结束。

下面演示利用 foreach 实现查询 pro_id 为 1~3 的产品信息。在 ProductMapper 类中添加如下 selectByRange 方法。

```
@Select("<script>" +
    "select * from product where pro_id in" +
    "<foreach item = 'id' index = 'index' collection = 'list' " +
    "open = '(' separator = ',' close = ')'>" +
    "#{id}" +
    "</foreach>" +
    "</script>")
List<Product> selectByRange(List<Integer> list);
```

在测试类 ProductMapperTest 中添加如下代码，调用 selectByRange 方法。

```
List<Integer> list=Stream.of(1,2,3).toList();
productMapper.selectByRange(list);
```

运行测试类，在控制台打印如下 SQL 信息，foreach 遍历了 list 集合 3 个元素并拼接进 SQL 语句中。

```
Preparing: select * from product where pro_id in ( ?, ?, ?)
Parameters: 1(Integer), 2(Integer), 3(Integer)
```

除了在查询场景中使用,foreach 还可以应用在批量数据更新场景中,只不过在大数据量或字段过多情况下有可能因拼接的 SQL 语句超出长度而执行失败,或因预编译时间过长而导致性能降低。因此 Mybatis 要求大数据量时批量数据更新操作使用 Batch 模式进行。

2. 基于@ * Provider 注解

上述基于 XML 动态标签需要在方法上写上大段的动态 SQL 语句,这很不简洁。因此,Mybatis 3.x 以上版本还提供@ * Provider 注解,允许自定义工具类以方法形式实现动态 SQL 语句。常用的@ * Provider 注解包含 @ InsertProvider、@ UpdateProvider、@ DeleteProvider 和@SelectProvider,对应数据的增、改、删和查操作。下面以 @ SelectProvider 为例介绍 @ * Provider注解语法。@SelectProvider 注解使用如下。

```
@SelectProvider(type = 类.class, method = "方法名")
```

其中主要包含两个属性 type 和 method,type 用于指定实现动态 SQL 语句的 Java 类,method 用于指定该类中实现动态 SQL 语句的具体方法。下面通过对 Product 表进行数据操作,演示@InsertProvider、@ UpdateProvider、@ DeleteProvider 和@SelectProvider 注解的使用。

在 springdemo5 目录下新建 provider 文件夹,文件夹内新建 ProviderProduct 类,类中定义 provider_insert、provider_update、provider_select 和 provider_delete 四个方法,用于对生成的数据进行增、改、查和删操作时的 SQL 语句。

【ProviderProduct.java】

```java
ProviderProduct.java
package springdemo5.provider;
import org.apache.ibatis.jdbc.SQL;
import springdemo5.domain.Product;
public class ProviderProduct {
    //新增数据
    public String provider_insert(Product product){
        return new SQL(){
            {
                INSERT_INTO("product");
                if(product.getPro_id()!=null){
                    VALUES("pro_id","#{pro_id}");
                }
                if(product.getPro_name()!=null){
                    VALUES("pro_name","#{pro_name}");
                }
            }
        }.toString();
    }
    //修改数据
    public String provider_update(Product product){
        return new SQL(){
            {
```

```
                UPDATE("product");
                if(product.getPro_id()!=null){
                    SET("pro_id=#{pro_id}");
                }
                if(product.getPro_name()!=null){
                    SET("pro_name=#{pro_name}");
                }
                WHERE("pro_id=#{pro_id}");
            }
        }.toString();
    }
    //查询数据
    public String provider_select(Product product){
        return new SQL(){
            {
                SELECT(" * ");
                FROM("product");
                if(product.getPro_id()!=null){
                    WHERE("pro_id=#{pro_id}");
                }
                if(product.getPro_name()!=null){
                    WHERE("pro_name=#{pro_name}");
                }
            }
        }.toString();
    }
    //删除数据
    public String provider_delete(Product product){
        return new SQL(){
            {
                DELETE_FROM("product");
                if(product.getPro_id()!=null){
                    WHERE("pro_id=#{pro_id}");
                }
                if(product.getPro_name()!=null){
                    WHERE("pro_name=#{pro_name}");
                }
            }
        }.toString();
    }
}
```

在 ProductMapper 类中新增 provider_insert、provider_update、provider_select 和 provider_delete 四个方法,分别使用@InsertProvider、@UpdateProvider、@DeleteProvider、@SelectProvider 和@DeleteProvider 修饰,用于引入不同的动态 SQL 语句。

```
@InsertProvider(type = ProviderProduct.class,method = "provider_insert")
@Options(useGeneratedKeys = true,keyProperty = "pro_id")//返回主键
int provider_insert(Product product);
```

```
@UpdateProvider(type = ProviderProduct.class,method = "provider_update")
int provider_update(Product product);
@SelectProvider(type = ProviderProduct.class,method = "provider_select")
List<Product>provider_select(Product product);
@DeleteProvider(type = ProviderProduct.class,method = "provider_delete")
int provider_delete(Product product);
```

在测试类 ProductMapperTest 中输入如下代码,调用 provider_insert 方法新增产品信息。

```
Product providerproduct=new Product();
providerproduct.setPro_name("product5");
productMapper.provider_insert(providerproduct);
System.out.println("返回主键:"+providerproduct.getPro_id());
```

运行测试类,执行成功,控制台输出如下信息,数据库中会看到如图 5-9 所示新增的数据。

```
Preparing: INSERT INTO product (pro_name) VALUES (?)
Parameters: product5(String)
Updates: 1
返回主键:4
```

在测试类 ProductMapperTest 中输入如下代码,调用 provider_update 方法,修改主键为 4 的 pro_name 为 product6。

```
productMapper.provider_update(new Product(4,"product6"));
```

运行测试类,执行成功,控制台输出如下信息,数据库中看到如图 5-10 所示修改后的数据。

```
Preparing: UPDATE product SET pro_id=?, pro_name=? WHERE (pro_id=?)
Parameters: 4(Integer), product6(String), 4(Integer)
Updates: 1
```

WHERE	
pro_id	pro_name
1	product1
2	product2
3	product4
4	product5

WHERE	
pro_id	pro_name
1	product1
2	product2
3	product4
4	product6

图 5-9　新增产品信息　　　　　　图 5-10　修改产品信息

在测试类 ProductMapperTest 中输入如下代码,调用 provider_select 方法,查询主键为 4 的 pro_name 为 product6 的产品。

```
System.out.println(productMapper.provider_select(
    new Product(4,"product6")));
```

运行测试类,执行成功,控制台输出如下信息。

```
Preparing: SELECT * FROM product WHERE (pro_id=? AND pro_name=?)
Parameters: 4(Integer), product6(String)
Total: 1
[Product(pro_id=4, pro_name=product6)]
```

在测试类 ProductMapperTest 中输入如下代码,调用 provider_delete 方法,删除主键为 4 的产品。

```
Product del_product=new Product();
del_product.setPro_id(4);
productMapper.provider_delete(del_product);
```

运行测试类,执行成功,控制台输出如下信息,数据库中看到主键为 4 的产品已经被删除了,如图 5-11 所示。

```
DELETE FROM product WHERE (pro_id=?)
Parameters: 4(Integer)
Updates: 1
```

图 5-11 删除产品信息

5.2.7 Mybatis 数据缓存机制

在企业级项目开发中,对于一些高频访问的数据查询结果,通常缓存在内存中,这样用户在查询数据的时候就直接访问缓存而不需要到数据库中查询,减少了用户与数据库的交互次数,提高了数据查询速度。尤其在高并发场景下,缓存的使用能够减少数据访压力,维护数据库性能稳定。

Mybatis 数据缓存机制

Mybatis 作为专业的数据访问持久层框架,也会提供相应的数据缓存机制,以优化数据访问。Mybatis 缓存总共分为两类:一级缓存和二级缓存。二者的区别在于作用域不同。前面提到 Mybatis 操作数据库需从会话工厂 SqlSessionFactory 中获取一条会话 SqlSession,通过 SqlSession 访问数据库。一级缓存是 SqlSession 级别的缓存,缓存只在当前 SqlSession 有效。而二级缓存是 SqlSessionFactory 级别的缓存,缓存可被多个 SqlSession 共享。

1. 一级缓存

一级缓存又称本地缓存,即在 SqlSession 中首次进行数据查询,结果会被缓存在内存中,随后利用该 SqlSession 再执行相同的数据查询操作(相同的数据查询操作指的是查询方法和传入参数都完全一致),就会直接从内存中获取数据,而不会再次访问数据库。Mybatis

默认已经开启了一级缓存。一级缓存只在当前 SqlSession 生命周期中生效，如果 SqlSession 关闭，则一级缓存失效。

下面以查询产品信息为例，演示一级缓存使用效果。这里不能像之前一样，在测试类中直接注入 mapper 接口类并两次执行相同的数据查询操作，因为这样每次查询都会开启一个新的 SqlSession，是不会触发一级缓存的。这里要利用同一个 SqlSession 两次执行相同的数据查询操作。在 springdemo5 的测试目录下新建 CacheTest.java 文件，内部注入 sqlSessionFactory 对象，利用 sqlSessionFactory 对象获取一个 SqlSession，并两次执行相同的数据查询操作。

【CacheTest.java】

```java
package springdemo5;
import org.apache.ibatis.session.SqlSession;
import org.apache.ibatis.session.SqlSessionFactory;
import org.junit.jupiter.api.Test;
import org.springframework.beans.factory.annotation.Autowired;
import org.springframework.boot.test.context.SpringBootTest;
import springdemo5.domain.Product;
import springdemo5.mapper.ProductMapper;
@SpringBootTest
public class CacheTest {
    //注入 sqlSessionFactory 对象
    @Autowired
    private SqlSessionFactory sqlSessionFactory;
    @Test
    void test(){
        //从 sqlSessionFactory 获取一条 sqlSession
        SqlSession sqlSession=sqlSessionFactory.openSession();
        ProductMapper productMapper= sqlSession.getMapper(ProductMapper.class);
        //第一次调用 selectProductByIdAndName
        productMapper.selectProductByIdAndName(new Product(1,"product1"));
        //第二次调用 selectProductByIdAndName
        productMapper.selectProductByIdAndName(new Product(1,"product1"));
        sqlSession.close();
    }
}
```

运行 CacheTest 的结果如下：程序总共只进行了一次 SQL 查询，第二次从一级缓存查询数据。

```
Preparing: select * from product where 1=1 and pro_id = ? and pro_name
Parameters: 1(Integer), product1(String)
Total: 1
```

如果想每次执行 selectProductByIdAndName 方法都从数据库获取最新数据，可以在 ProductMapper 类的 selectProductByIdAndName 方法上添加如下 Options 注解。

```
@Options(useCache = true, flushCache = Options.FlushCachePolicy.TRUE)
```

在每次执行该方法时都清空之前的缓存,从数据库中查询数据。其中 flushCache 值为 Options.FlushCachePolicy 的枚举值。如果值为 TRUE,每次查询时清空缓存;如果值为 FALSE,每次查询从缓存中获取数据。

如果在 selectProductByIdAndName 方法上已经添加了以上的@Options 注解,再次运行 CacheTest,结果如下:两次查询都清空缓存并从数据库查询。如果设置 flushCache = Options.FlushCachePolicy.FALSE,则只有第一次查询使用数据库,第二次查询使用一级缓存。

```
Preparing: select * from product where 1=1 and pro_id = ? and pro_name
Parameters: 1(Integer), product1(String)
Total: 1
Preparing: select * from product where 1=1 and pro_id = ? and pro_name
Parameters: 1(Integer), product1(String)
Total: 1
```

此外,如果出现以下三种情况,程序也会不使用一级缓存。

(1) 使用不同的 SqlSession 进行两次相同的数据查询操作,不会使用原有 SqlSession 的一级缓存。例如,如下代码将进行两次数据库查询。

```
//获取两个不同的 sqlSession 对象
SqlSession sqlSession=sqlSessionFactory.openSession(true);
SqlSession sqlSession1=sqlSessionFactory.openSession(true);
ProductMapper productMapper= sqlSession.getMapper(ProductMapper.class);
ProductMapper productMapper1= sqlSession1.getMapper(ProductMapper.class);
//第一次调用 selectProductByIdAndName
productMapper.selectProductByIdAndName(new Product(1,"product1"));
//第二次调用 selectProductByIdAndName
productMapper1.selectProductByIdAndName(new Product(1,"product1"));
sqlSession.close();
```

(2) 使用同一个 SqlSession 进行相同的数据查询操作,但两次查询中间执行了数据增、改和删操作,一级缓存数据会被清空。例如,如下代码将进行两次数据库查询。

```
SqlSession sqlSession=sqlSessionFactory.openSession(true);
ProductMapper productMapper= sqlSession.getMapper(ProductMapper.class);
//第一次调用 selectProductByIdAndName
productMapper.selectProductByIdAndName(new Product(1,"product1"));
//两次查询中间进行数据修改操作
productMapper.updateProduct(new Product(2,"product3"));
//第二次调用 selectProductByIdAndName
productMapper.selectProductByIdAndName(new Product(1,"product1"));
sqlSession.close();
```

(3) 使用同一个 SqlSession 进行相同的数据查询操作,但是两次查询中间使用 SqlSession 的 clearCache 方法手动清空缓存,一级缓存数据会被清空。例如,如下代码将进行两次数据库查询。

```
//从 sqlSessionFactory 中获取一条 sqlSession
SqlSession sqlSession=sqlSessionFactory.openSession(true);
ProductMapper productMapper= sqlSession.getMapper(ProductMapper.class);
//第一次调用 selectProductByIdAndName
productMapper.selectProductByIdAndName(new Product(1,"product1"));
//第二次调用 selectProductByIdAndName
sqlSession.clearCache();
productMapper.selectProductByIdAndName(new Product(2,"product2"));
sqlSession.close();
```

2. 二级缓存

二级缓存又称为全局缓存,是 SqlSessionFactory 级别的缓存,可以被同一 SqlSessionFactory 创建的多个 SqlSession 共享,缓存数据存储在 SqlSessionFactory 中。当该 sqlSession 中没有数据增、改和删操作,执行 sqlSession.close()或 sqlSession.commit(),才会将该 sqlSession 一级缓存中的数据写入二级缓存。由于二级缓存数据既可能存在内存中,又可能存在磁盘中,缓存数据取出需执行反序列化操作,因此二级缓存要求实体类必须实现 Serializable 接口。

Mybatis 默认关闭二级缓存,使用时需手动开启,并基于 mapper 接口类进行配置,同一个 mapper 接口类中的所有数据操作方法共享一个二级缓存,是表级别的数据缓存。如果开启二级缓存,数据查询顺序为先查询二级缓存,如果没有再查询一级缓存,也不会查询数据库。在同一个 mapper 类的两次查询之间对该 mapper 类映射的表数据执行了数据增、改和删操作,且执行 sqlSession.close()或 sqlSession.commit()操作,二级缓存才会被清空。否则二级缓存默认使用 LRU 原则清除,即清除 SqlSessionFactory 中最长时间不被使用的数据对象。

在使用时,对于数据查询比较多的项目可以考虑配置二级缓存,Spring Boot 已经简化了 Mybatis 二级缓存的配置和使用,只需进行三步配置即可使用。

① 在 application.yaml 配置中开启全局二级缓存。
② 在相应的实体类中实现 Serializable 接口。
③ 在相应的 mapper 类上添加@CacheNamespace 注解。

下面以查询产品信息为例演示二级缓存的用法。

(1) 在 application.yaml 配置开启全局二级缓存。

```
mybatis:
  configuration:
    cache-enabled: true
```

(2) 在相应的实体类实现 Serializable 接口。

```
public class Product implements Serializable
```

(3) 在相应的 mapper 类上添加@CacheNamespace 注解。

```
@CacheNamespace
public interface ProductMapper {
```

```
    //省略内部方法
}
```

(4) 修改测试类 CacheTest 的 test 方法,内部通过三个不同的 SqlSession 调用 3 次数据查询方法。

```
void test(){
    //从 sqlSessionFactory 获取三条不同的 sqlSession
    SqlSession sqlSession1=sqlSessionFactory.openSession(true);
    SqlSession sqlSession2=sqlSessionFactory.openSession(true);
    SqlSession sqlSession3=sqlSessionFactory.openSession(true);
    ProductMapper productMapper1= sqlSession1.getMapper(ProductMapper.class);
    ProductMapper productMapper2= sqlSession2.getMapper(ProductMapper.class);
    ProductMapper productMapper3= sqlSession3.getMapper(ProductMapper.class);
    //第一次调用 selectProductByIdAndName
    productMapper1.selectProductByIdAndName(new Product(1,"product1"));
    //在 sqlSession1 没有数据修改操作前提下,调用 sqlSession1 的 commit 方法或 close
    //方法,二级缓存才存入数据
    sqlSession1.commit();
    //第二次调用 selectProductByIdAndName
    productMapper2.selectProductByIdAndName(new Product(1,"product1"));
    //第三次调用 selectProductByIdAndName
    productMapper3.selectProductByIdAndName(new Product(1,"product1"));
}
```

运行 CacheTest 结果如下:其中只进行了一次数据库查询,后面两次查询都使用二级缓存。Cache Hit Ratio 缓存命中率这个指标显示了 ProductMapper 接口类二级缓存的命中情况,第一次查询从数据库查询,命中情况为 0。第二次查询使用二级缓存,命中情况为 1/2=0.5。第三次查询也使用二级缓存,命中情况为 2/3=0.66。

```
Cache Hit Ratio [springdemo5.mapper.ProductMapper]: 0.0
Preparing: select * from product where 1=1 and pro_id = ? and pro_name = ?
Parameters: 1(Integer), product1(String)
Total: 1
Cache Hit Ratio [springdemo5.mapper.ProductMapper]: 0.5
Cache Hit Ratio [springdemo5.mapper.ProductMapper]: 0.6666666666666666
```

Mybatis 的二级缓存相对于一级缓存来说,实现了 SqlSession 之间共享缓存的数据,将数据缓存细化到了 mapper 级别。

任务 5.3　Spring Boot 事务

在数据增、删、改过程中,事务控制必须不少。如银行卡转账业务,如果 A 账户转给 B 账户过程中出错,不进行事务控制很有可能 A 的钱减少,B 账户的钱没增加,这样会造成经济损失。如果进行了事务控制,就可以避免上述问题。本任务将介绍在数据增、删、改过程中如何使用 Spring Boot 进行事务控制。

5.3.1 事务简介

事务指的是一组数据操作组成的一个逻辑单元,该逻辑单元内的数据操作要么全部执行成功,要么全部不执行。事务具有以下四个特征。

原子性:事务是一个不可分割的工作单位,事务中的操作要么都发生,要么都不发生。

一致性:事务执行前后数据的完整性必须保持一致。

隔离性:当有多个事务并发访问数据库时,数据库为每一个访问开启事务,各并发访问的事务之间相互隔离。

持久性:事务一旦被提交,它对数据库中数据的改变就是永久性的。

事务的作用就是在数据增、删、改操作时,保证数据库数据的一致性和可靠性。如某业务的数据修改涉及多个数据表,如果其中一个数据表的数据修改完毕出现异常,后续其他数据表的数据修改操作将无法执行,就会破坏整个数据库数据的一致性,这就需要使用事务管理来确保所有数据修改操作在同一事务中执行,要么全部执行,要么全部不执行,保证了整个数据库数据的一致性。又如在数据库并发访问中,访问 A 和 B 时时读入同一数据并修改,如果没有事务控制,访问 A 的数据修改结果可能会覆盖了访问 B 的数据修改结果,导致访问 B 数据修改结果丢失,造成数据异常。

5.3.2 Spring Boot 声明式事务控制

Spring Boot 和 Mybatis 整合以后,Mybatis 事务控制交给 Spring Boot 管理。Spring Boot 的事务控制基于 Spring。Spring 支持声明式事务,以注解方式实现事务控制。Spring 使用声明式事务只需做如下两步操作。

(1) 使用注解@EnableTransactionManagement 在 Spring 配置类上开启声明式事务。

(2) 在需要进行事务控制的 Java 类或方法上添加@Transactional 注解,当@Transactional 使用在类上,此类的所有 public 方法都进行事务控制。当@Transactional 使用在方法上,只有当前方法进行事务控制,方法内部的所有数据操作成为一个事务整体,共享一个数据库连接,如果方法内部发生了 error 或 RuntimeException 异常并抛出,Spring 会利用 AOP 机制捕获异常,所有数据操作自动回滚。如果@Transactional 同时使用在类和方法上,方法上的@Transactional 配置将覆盖类上的配置。

Spring Boot 默认注入一个事务管理器并自动开启声明式事务,因此只需进行第二步操作,在进行事务控制的 Java 类或方法上添加@Transactional,即可进行事务控制。@Transactional 注解内部定义了多个属性,@Transactional 内部属性如表 5-2 所示。

表 5-2 @Transactional 内部属性

属性	描述
value	指定使用的事务管理器,别名为 transactionManager,默认为""
transactionManager	别名为 value,与 value 作用类似,用于指定使用的事务管理器,默认为""
isolation	指定事务隔离级别,默认为 Isolation.DEFAULT
propagation	指定事务的传播行为,默认为 Propagation.REQUIRED

续表

属性	描述
timeout	指定事务超时情况,默认为-1,表示永不超时
readOnly	指定是否只读,默认为 false
rollbackFor	指定发生哪种异常进行事务回滚,值为异常类型,有多个异常类型时用逗号分隔
rollbackForClassName	指定发生多种异常时进行事务回滚,值为异常类名的字符串数组
noRollbackFor	指定发生哪种异常时不进行事务回滚,值为异常类型,有多个异常类型时用逗号分隔
noRollbackForClassName	指定发生多种异常时不进行事务回滚,值为异常类名的字符串数组

5.3.3 事务隔离级别

在实际场景中,对数据库大多是并发访问,每个访问都基于各自的事务,事务之间难免产生干扰。这就需要用到@Transactional 的 isolation 属性,isolation 通过设置事务隔离级别解决并发事务访问之间的数据干扰问题,包括数据丢失、脏读、幻读、不可重复读等问题。

数据丢失:两个事务 T1 和 T2 读入同一数据并修改,T2 的提交结果覆盖了 T1 提交的结果,导致 T1 的修改数据丢失。

脏读:事务 T1 修改某数据后,事务 T2 读取该数据。T1 由于异常被操作回滚,数据恢复原值。T2 读到的数据值还是回滚之前的值,T2 读到的数据就是脏数据。

幻读:事务 T1 对表做多记录查询,事务 T2 对表中新增或删除一条数据。事务 T1 再次查询出来的数据和之前查询的数据不一致。

不可重复读:事务 T1 查询单条数据记录,事务 T2 对其做了修改。事务 T1 再次查询该数据时和之前查询的数据不一致。

Spring 内部定义了如表 5-3 所示的五种事务隔离级别,以解决上述问题。

表 5-3 事务隔离级别

隔离级别	值	描述
ISOLATION_DEFAULT	DEFAULT(-1)	Spring 默认的隔离级别,使用当前数据库的隔离级别,大部分数据库为 READ_COMMITTED
ISOLATION_READ_UNCOMMITTED	READ_UNCOMMITTED(1)	读未提交:事务可以读取另一个事务修改但还没有提交的数据。解决了数据丢失问题,没有解决脏读、不可重复读和幻读问题
ISOLATION_READ_COMMITTED	READ_COMMITTED(2)	读已提交:事务可读取另一个事务修改且提交的数据。解决了数据丢失和脏读问题。没有解决数据不可重复读和幻读问题

续表

隔离级别	值	描述
ISOLATION_REPEATABLE_READ	REPEATABLE_READ(4)	重复读：该隔离级别是 MySQL 的默认隔离级别。事务在整个过程中可以多次重复执行某查询，并且每次返回的记录都相同。即使在多次查询之间有数据修改操作，也会被忽略。解决了更新丢失、脏读和不可重复问题，没有解决数据幻读问题
ISOLATION_SERIALIZABLE	SERIALIZABLE(8)	串行化：以串行方式依次逐个执行所有事务，事务之间完全不干扰，解决了事务并发的所有问题，但是会牺牲数据库性能

5.3.4 事务传播机制

在编程中，service 层的业务逻辑方法往往会相互调用，但每个业务逻辑方法都有自己的事务，那么这些不同的事务之间该如何调度执行？这就需要了解事务的传播机制，即 @Transactional 的 propagation 属性。Spring 内部定义了如表 5-4 所示的七种事务传播机制。

表 5-4 事务传播机制

传播机制	值	描述
PROPAGATION_REQUIRED	REQUIRED(0),	Spring 默认传播机制，若 A 调用 B 方法，A 存在事务则 B 使用 A 事务，A 没有事务则 B 新建事务
PROPAGATION_SUPPORTS	SUPPORTS(1)	若 A 调用 B 方法，A 存在事务则 B 使用 A 事务，A 没有事务则 B 以非事务方式执行
PROPAGATION_MANDATORY	MANDATORY(2)	若 A 调用 B 方法，B 方法只能在事务中执行，A 存在事务则 B 使用 A 事务，A 没有事务则抛出异常
PROPAGATION_REQUIRES_NEW	REQUIRES_NEW(3)	若 A 调用 B 方法，A 存在事务则挂起 A 事务，新建一个新事务执行 B，执行完毕，恢复挂起的 A 事务
PROPAGATION_NOT_SUPPORTED	NOT_SUPPORTED(4)	若 A 调用 B 方法，B 方法只能以非事务方式执行，如 A 有事务则挂起，待 B 执行完毕，恢复 A 事务
PROPAGATION_NEVER	NEVER(5)	若 A 调用 B 方法，B 方法只能以非事务方式执行，如 A 有事务则抛出异常
PROPAGATION_NESTED	NESTED(6)	若 A 调用 B 方法，无论 A 是否则存在事务，都新建一个嵌套事务执行 B

5.3.5 编程实现基于注解的事务控制

下面结合编程演示事务控制的效果，在 springdemo5 目录下新建 service 文件夹，内部新建 ProductService 类。ProductService 类上添加@Transactional 注解进行事务控制，内部定义的业务方法可调用 ProductMapper 类中的 insertProduct 方法对产品信息表插入一条记录，人为抛出一个 RuntimeException 类异常，让插入操作自动回滚。

【ProductService.java】

```
package springdemo5.service;
import org.springframework.beans.factory.annotation.Autowired;
import org.springframework.stereotype.Service;
import org.springframework.transaction.annotation.Transactional;
import springdemo5.domain.Product;
import springdemo5.mapper.ProductMapper;
@Transactional
@Service
public class ProductService {
    @Autowired
    private ProductMapper productMapper;
    public void insertProduct(){
        Product product=new Product(5,"product5");
        productMapper.insertProduct(product);
        throw new RuntimeException("error");
    }
}
```

创建 ProductService 的测试类 ProductServiceTest，在测试类中调用 insertProduct 方法。

【ProductServiceTest.java】

```
package springdemo5.service;
import org.junit.jupiter.api.Test;
import org.springframework.beans.factory.annotation.Autowired;
import org.springframework.boot.test.context.SpringBootTest;
@SpringBootTest
class ProductServiceTest {
    @Autowired
    private ProductService productService;
    @Test
    void insertProduct() {
        productService.insertProduct();
    }
}
```

运行测试类，能够看到控制台输出如下信息。证明程序确实执行了插入操作。但是数据库数据如图 5-12 所示，并没有新增数据，这是因为事务控制生效，业务方法抛出了 RuntimeException 异常，插入操作被自动回滚了。

pro_id	pro_name
1	product1
2	product2
3	product4

图 5-12 插入操作自动回滚

```
Preparing: insert into product ( pro_id, pro_name ) values ( ?, ?)
Parameters: 5(Integer), product5(String)
Updates: 1
java.lang.RuntimeException: error
    at springdemo5.service.ProductService.insertProduct(ProductService.java:16)
```

在编程中,如果添加了@Transactional 注解,但异常发生事务没有自动回滚,可从以下几个方面查找原因。

(1) 事务控制的方法必须是一个公共方法,被 public 修饰,否则事务不自动回滚。

(2) 事务控制的方法中异常必须被抛出,无论是直接抛出还是使用"try{ } catch{throw}"抛出都可以,否则事务不自动回滚。如不想抛出异常,也要事务回滚,可添加如下代码来手动回滚事务。

```
TransactionAspectSupport.currentTransactionStatus().setRollbackOnly();
```

(3) 事务控制的方法中抛出的异常必须是 error 或 Exception 的子类 RuntimeException,其他类的异常事务不自动回滚。如确实要对所有 Exception 类异常自动回滚,可以在业务类或方法上添加@Transactional(rollbackFor = {Exception.class})。

(4) 事务控制的方法如果内部开启新线程,则不同线程的数据库连接不同,事务不自动回滚。

(5) 数据库需支持事务回滚,如 MySQL 数据库引擎为 InnoDB 才支持事务回滚。

任务 5.4　综合案例:用 Spring Boot 模拟实现人员账户管理

基于上述知识,本节以一个综合案例演示 Spring Boot+Mybatis 数据操作的编程应用,案例以人员账户管理为例,编程中需要用到数据库连接池、Mybatis 关联映射、动态 SQL、事务控制和缓存等知识点。

5.4.1　案例任务

在数据库中创建用户表(users)和账户表(accounts)。users 表字段为用户 id(user_id)、用户名(user_name)、电话(phone),user_id 为主键。accounts 表字段为账户 id(account_id)、金额(money)、创建时间(create_time)、账户密码(account_key)、账户所属用户(u_id),account_id 为主键,u_id 为外键,u_id 的值参考 user_id。users 表和 accounts 表为一对多关系,一个人可以有多个账户。编程实现如下四个功能。

(1) 开户功能:注册新用户并新开一个账户。

(2) 查询功能:用户可以查询自己的所有账户信息。

(3) 修改功能:用户可以修改自己的电话号码和设置某个账户的密码。

(4) 销户功能:删除用户并删除账户。

其中,数据库连接池使用 Druid,数据操作使用动态 SQL 并根据需要添加事务控制。

5.4.2 案例分析

该任务涉及单表数据操作、一对多关联查询、动态 SQL、事务控制多表数据修改,以及数据库连接池 Druid 配置等知识点。该任务基于 SpringDemo5 项目实现,相关配置使用原有配置。任务的实现思路如下。

(1) 新建数据库并创建 users 和 accounts 表,分别向两张表中添加数据。
(2) 创建实体类 Users 和 Accounts。
(3) 创建 Users 和 Accounts 的 mapper 映射接口类,内部使用动态 SQL 编写相应的数据增、删、改、查方法,映射接口类上添加的 @CacheNamespace。
(4) 创建业务类并注入 mapper 映射接口,内部根据业务需求组合 mapper 类的方法实现业务功能,根据需要添加事务控制。
(5) 编写测试类测试结果。

5.4.3 任务实施

由于篇幅有限,任务实施具体步骤采用二维码形式展示。

综合案例的实施步骤

小 结

本项目详细介绍了基于 MySQL 的 Spring Boot 数据操作和事务处理,包括数据库连接池的使用、Mybatis 的使用(单表查询、关联查询、动态 SQL 和缓存)、Spring 声明式事务控制等内容。最后以一个"综合案例:用 Spring Boot 模拟实现人员账户管理"演示如何在 Spring Boot 项目中综合运用数据库连接池、Mybatis 关联查询、动态 SQL 和缓存等知识点完成简单的业务逻辑功能,使读者进一步掌握 Spring Boot 的数据读写编程。

课后练习:用 Spring Boot 模拟实现人员账户转账

进一步完善综合案例项目,在本项目的任务 4 基础上编程实现如下两个业务功能。
(1) 账户存取功能:用户可以向自己的账户存钱和取钱。
(2) 账户转账功能:用户可以用自己的多个账户互相转账,也可以转账给他人账户。

项目 6　Spring Boot 定时任务

在开发中经常会有这样的业务场景需求,如系统在每天晚上 12 点进行当天数据备份,系统运行时每小时写一次日志文件,系统每隔 10 分钟定时爬取门户网站的更新数据,系统每天早上 7 点准时推送最新新闻等。这些场景往往都要求我们在某个特定的时间去做某件事情,完成一些周期性的任务。这时就需要用到定时任务框架来进行编程。本项目主要介绍如何用 Spring Boot 整合定时任务框架 SpringTask 来编程实现定时任务。

任务 6.1　Cron 表达式和定时任务框架

编程实现定时任务,最简单的方式就是设置任务执行的绝对时间或时间间隔。如设定任务在晚上 12:00 开始执行或设定任务启动后每隔 5 秒执行一次。在编程中一般使用 Cron 表达式来设置任务定时执行时间。Cron 表达式是 Linux 系统中的定时任务执行工具,可以在无须人工干预的情况下执行任务。同时 Cron 表达式也可方便地定义一些更为复杂的任务执行策略。本任务中主要介绍 Cron 表达式以及一些常见的定时任务调度框架。

6.1.1　初识 Cron 表达式

Cron 表达式

在 Linux 系统中经常会使用 Cron 表达式来定义定时任务的执行策略。Cron 表达式又称时间表达式,是一个字符串。该字符串以 5 或 6 个空格隔开,被分为 6 个或 7 个域,每个域都代表一个时间维度。Cron 表达式具体格式如下。

```
//从左到右分别表示:秒 分 时 日 月 周 年,参数以空格隔开
//最后一个参数年为非必需,可以省略。省略年为 6 个域,不省略年为 7 个域
{Seconds} {Minutes} {Hours} {DayofMonth} {Month} {DayofWeek} {Year}
```

Cron 的各域的定义如表 6-1 所示。

表 6-1　Cron 各域的定义

域	是否必须	参 数 范 围	可使用的通配符
秒	是	0～59	, - * /
分	是	0～59	, - * /
时	是	0～23	, - * /
日	是	1～31	, - * ? / L W
月	是	1～12 或 JAN 至 DEC	, - * /
周（周一至周日）	是	1～7(1＝SUN) 或 SUN,MON,TUE,WED,THU,FRI,SAT	, - * ? / L #
年	否	1970～2099	, - * /

通配符的应用是 Cron 表达式中的一个难点，下面对表中通配符进行介绍。

(1) ,：以枚举形式列举任务执行的多个时间，如在 Minutes 域中定义为"1,2"，则表示分别在第 1 分钟和第 2 分钟执行该定时任务。

(2) -：表示一段连续时间范围，如在 Minutes 域中定义 10～15，则表示在 10～15 分钟每隔 1 分钟执行 1 次任务，当然也可以用逗号枚举表示。

(3) *：表示匹配任意值。假如在 Minutes 域定义 *，表示每分钟都执行 1 次任务。

(4) ?："?"只能用在 DayofMonth 和 DayofWeek 两个域，表示不指定值，两个域不能同时使用"?"。例如在每月的 20 日执行任务，不指定周几，可使用如下写法"0 0 20 * ?"。

(5) /：表示间隔时间执行任务。如在 Minutes 域上定义 0/5，表示从 0 分开始，每隔 5 分钟执行一次任务。

(6) L：只能用在 DayofWeek 和 DayofMonth 域，表示最后。如果在 DayofMonth 域中定义，表示每个月的最后一天执行任务。如果在 DayofWeek 域中定义表示周六，相当于 7 或 SAT。如果在 L 前加上数字，表示最后一个周几执行任务。如在 DayofWeek 域定义 4L，表示最后一个周三执行任务。

(7) W：只能用在 DayofMonth 域，表示寻找当月离指定日期最近的工作日（周一至周五）执行任务。例如在 DayofMonth 域定义 10W，表示离每月 10 号最近的那个工作日执行任务。假如 10 号正好是周六，则找本周五（9 号）执行任务；如果 10 号是周日，则找下周一（11 号）执行任务；如果 10 号正好是工作日，则就在 10 号当天执行任务。

(8) ♯：只能用在 DayofWeek 域，表示每月的第几个周几。如 4♯3 表示在每月的第三个周三。

下面列举一些通配符的案例写法。

```
cron ="0 0 0 * * ?"      每天零点执行一次
cron ="0 */10 * * * ?"    每隔 10 分钟执行一次
cron ="0 0/10 14 * * ?"   在每天下午 2 点到下午 2:50 的每 10 分钟执行一次
cron ="0 0/10 14,18 * * ?"  在每天下午 2:00—2:50 和 6:00—6:50 每 10 分钟执行一次
cron ="0 0-5 14 * * ?"   在每天下午 2 点到下午 2:05 每隔 1 分钟执行一次
cron ="0 0 13 ? * 1"     每周日下午 1 点执行一次
cron ="0 30 8 ? * 5#2"   每月的第二个星期四上午 8:30 执行一次
cron ="0 10,20 12 ? 3 1"  每年三月周日中午 12:10 和 12:20 执行一次
cron ="0 30 8 L * ?"     每月最后一天的上午 8:30 执行一次
cron ="0 30 8 ? * 4L"    每月最后一个周三上午 8:30 执行一次
```

6.1.2　常用的定时任务框架

目前常见的定时任务框架主要有以下几种。

（1）Timer：Timer 是 JDK 自带的定时器类，内部执行 TimerTask 类型任务。支持特定时间点执行任务，不支持 Cron 设置任务执行时间。任务只能单线程执行，前面任务执行时间过长会影响后续任务的执行。发生异常时后续任务直接停止，无法执行。

（2）ScheduledExecutorService：ScheduledExecutorService 也是 JDK 自带的定时器接口，常用的实现类是 ScheduledThreadPoolExecutor，ScheduledExecutorService 的出现用于替代 Timer。ScheduledExecutorService 是基于线程池的多线程任务执行。ScheduledExecutorService 支持以相对延迟或者周期设置任务执行时间，不支持 Cron 设置任务执行时间。发生异常时后续任务仍然可以执行。

（3）Spring Task：Spring 自带的一款主流任务调度框架，配置简单且使用方便，专门用于定时任务执行，功能比较强大，支持以相对延迟或者周期设置任务执行时间，也支持 Cron 设置任务执行时间。Spring Task 底层使用 JDK 的 ScheduledThreadPoolExecutor 单线程线程池调度任务，任务执行采用同步方式，多任务需排队执行。Spring Task 不支持分布式任务，是单机环境下定时任务框架的首选。

（4）Quartz：Quartz 是市面上所有定时任务框架的鼻祖，功能强大，可与 Spring 集成，支持 Cron 设置任务执行时间。Quartz 无可视化管理界面，相比 Spring Task 使用相对复杂。Quartz 基于数据库实现集群分布式任务调度，集群扩容能力受数据库性能影响，可用于单机环境定时任务，也可用于中小规模集群定时任务。

（5）Elastic-Job：一个基于 Quartz 和 ZooKeeper 的开源分布式调度框架，功能强大。可与 Spring 集成，支持以相对延迟或者周期设置任务执行时间，也支持 Cron 设置任务执行时间，有可视化管理界面。使用时需整合 ZooKeeper，学习成本相对复杂。Elastic-Job 基于 ZooKeeper 实现集群分布式任务调度，集群扩容能力强，一般用于大规模集群任务调度。

（6）XXL-Job：一款开源的轻量级任务调度框架，可与 Spring 集成，支持以相对延迟或者周期设置任务执行时间，也支持 Cron 设置任务执行时间，学习成本低，使用方便，有可视化管理界面。XXL-JOB 使用了 Quartz 基于数据库实现集群分布式任务调度，集群扩容能力受数据库性能影响，一般用于中小规模集群任务调度。

下面对上述框架的关键特征做综合对比，如表 6-2 所示。

表 6-2　定时任务框架

名称	分布式	并行任务	扩展性	Cron	使用场景
Timer	不支持	不支持	无	不支持	单机环境单线程任务
ScheduledExecutorService	不支持	支持	无	不支持	单机环境多线程任务
Spring Task	不支持	支持	无	支持	单机环境单/多线程任务
Quartz	支持	支持	较强	支持	中小规模集群任务调度
Elastic-Job	支持	支持	强	支持	大规模集群任务调度
XXL-Job	支持	支持	较强	支持	中小规模集群任务调度

任务 6.2　基于 Spring Task 定时任务编程

Spring Task 是 Spring 框架的一部分，用于实现定时任务调度。本任务主要介绍如何在 Spring Boot 项目中使用 Spring Task 编程实现定时任务，包括单任务和多任务的编程。

6.2.1　初识 Spring Task

Spring Task 是 spring-context 模块下提供的定时任务工具。Spring Task 使用起来更简单，可以看作是一个轻量级的 Quartz，是单机环境下定时任务框架的首选。Spring Task 默认采用单线程任务调度，任务同步执行机制来执行定时任务。使用 Spring Task 不需添加其他依赖，只需在项目中导入 spring-context 依赖即可。Spring Task 不支持分布式任务，但是能够通过其他方式实现分布式任务调度。

使用 Spring Task 有两种方式：一种是基于注解，另一种是基于配置文件。这里主要基于注解介绍 Spring Task 的使用。基于注解方式 Spring Task 的运行原理如下。

（1）通过后置处理器 ScheduledAnnotationBeanPostProcessor 监听 Spring IOC 容器初始化事件 ContextRefreshedEvent。

（2）待 Spring IOC 容器初始化完毕，扫描所有 Bean 中带有 @Scheduled 注解的方法，然后封装成任务计划 ScheduledTask 注册并到注册中心 ScheduledTaskRegistrar 中等待执行。

（3）ScheduledTaskRegistrar 内部通过 Executors.newSingleThreadScheduledExecutor 方法创建一个单线程线程池的定时任务执行器 ScheduledThreadPoolExecutor(1)，并注入并发任务调度器 ConcurrentTaskScheduler 中。

（4）通过并发任务调度器 ConcurrentTaskScheduler 执行定时任务，如有多个定时任务需排队执行。同时 ScheduledTaskRegistrar 也支持创建线程池以多线程方式执行定时任务。

6.2.2　Spring Task 基于单个定时任务编程实现

在 Spring Boot 项目中使用 Spring Task，只需在 pom.xml 文件中引入 spring-context 依赖即可，这里为了方便直接在 pom.xml 文件中引入 spring-webmvc 依赖，spring-context 会被自动引入。引入后 Spring Boot 会对 Spring Task 自动配置，可直接使用。

基于注解的 Spring Task 编程，主要利用 @EnableScheduling 和 @Scheduled 两个注解。@EnableScheduling 注解用于开启定时任务，添加在需要开启定时任务的类上。在 Spring Boot 项目中，@EnableScheduling 一般添加在 Spring Boot 启动类上。@Scheduled 用于执行定时任务的具体方法上设置定时任务的执行逻辑。@Scheduled 注解有四个常用属性用于设置定时任务，如表 6-3 所示。

表 6-3 @Scheduled 的四个常用属性

名称	描述	使用方法
fixedDelay	设定上个任务结束后多久执行下个任务，时间单位为毫秒	@Scheduled(fixedDelay=1000)，表示上个任务完成后间隔 1000 毫秒再执行下个任务。由于任务同步执行，1000 毫秒内上个任务没有执行完，会等待上个任务执行完毕再延迟 1000 毫秒开始执行下个任务。假设此时任务执行时间为 5000 毫秒，大于 fixedDelay 值，则 fixedDelay 失效，两次任务执行时间间隔为 5000+1000=6000（毫秒）
fixedRate	设定上个任务开始后多久执行下个任务，时间单位为毫秒	@Scheduled(fixedRate=1000)，表示上个任务开始后 1000 毫秒执行下个任务。由于任务同步执行，1000 毫秒内上个任务没有执行完，会等待上个任务执行完毕才开始执行下个任务。假设此时任务执行时间为 5000 毫秒，大于 fixedRate 值，则 fixedRate 失效，两次任务执行时间间隔为 5000 毫秒
initialDelay	设定延迟多长时间后开始执行第一次定时任务。一般配合 fixedDelay 或 fixedRate 使用	@Scheduled(initialDelay=1000, fixedRate=2000)，表示任务初始执行延迟 1000 毫秒，每隔 2000 毫秒执行下次任务
cron	以 cron 表达式设置定时任务	@Scheduled(cron = "0/1 * * * * ?")，表示每 1000 毫秒执行一次任务。如果任务执行时间为 5000 毫秒，大于 cron 值，两次任务执行时间间隔为 5000 毫秒，结果与 fixedRate 类似

其中，fixedRate 和 fixedDelay 主要用于设置简单定时任务，cron 除了可以设置简单定时任务，还可用于设置复杂定时任务。下面以执行单个定时任务为例演示这四种属性的使用。在 Idea 中新建名为 SpringDmoe6 的 Spring Boot 项目，在项目的 pom.xml 文件中添加如下 spring-webmvc 依赖。

```
<dependency>
    <groupId>org.springframework.boot</groupId>
    <artifactId>spring-boot-starter-web</artifactId>
</dependency>
<dependency>
    <groupId>org.springframework.boot</groupId>
    <artifactId>spring-boot-starter-test</artifactId>
    <scope>test</scope>
</dependency>
```

在项目启动类 Springdemo6Application 上添加注解 @EnableScheduling，开启定时任务。在项目的 springdemo6 目录下新建 task 目录。在 task 目录下新建 ThreadTask 类，内部定义的 simpleTask 方法打印当前时间。simpleTask 方法使用注解 @Scheduled (initialDelay = 1000, fixedDelay = 2000)，设置延迟 1000 毫秒开始执行第一次任务，待第一次任务完成后间隔 2000 毫秒再执行下一次任务。

```
package springdemo6.task;
import org.springframework.scheduling.annotation.Scheduled;
import org.springframework.stereotype.Component;
import java.time.LocalTime;
```

```
@Component
public class ThreadTask {
    @Scheduled(initialDelay = 10000,fixedDelay = 2000)
    public void simpleTask() throws InterruptedException {
        String threadName=Thread.currentThread().getName();
        System.out.println(threadName+"--simpleTask process:" + LocalTime.now());
    }
}
```

运行 Springdemo6Application 类,项目启动 10000 毫秒后,控制台输出如下信息,simpleTask 方法使用单线程 scheduling-1 执行定时任务。首次任务延迟 1000 毫秒执行,后续任务执行时间间隔为 2000 毫秒。

```
scheduling-1--simpleTask process:16:02:13.003034100
scheduling-1--simpleTask process:16:02:15.005393500
scheduling-1--simpleTask process:16:02:17.008013600
```

如果将 @Scheduled(initialDelay = 10000,fixedDelay = 2000) 改成 @Scheduled(initialDelay = 10000,fixedRate = 2000)或@Scheduled(cron = "0/2 * * * * ?"),由于 simpleTask 执行时间很短,也可以得到以上结果,任务执行时间间隔为 2000 毫秒。

下面修改 simpleTask,让 simpleTask 休眠 5000 毫秒,以增加任务执行时间。由于任务同步执行,这时 fixedDelay 和 fixedRate 将呈现不同结果,而 cron 和 fixedRate 结果相同。这里先设置 fixedDelay = 2000。

```
@Scheduled(initialDelay = 10000,fixedDelay = 2000)
public void simpleTask() throws InterruptedException {
    String threadName=Thread.currentThread().getName();
    System.out.println(threadName+"--simpleTask process:" + LocalTime.now());
    Thread.sleep(5000);         //模拟长时间执行,休眠 5000 毫秒
}
```

运行 Springdemo6Application 类,项目启动 10000 毫秒后,控制台输出如下信息。任务同步执行时,fixedDelay = 2000 表示在上一次任务执行完毕再间隔 2000 毫秒执行下次任务,而 simpleTask 运行时间为 5000 毫秒,所以 simpletask 两次任务运行时间间隔为 5000+2000=7000(毫秒)。

```
scheduling-1--simpleTask process:16:17:37.307816500
scheduling-1--simpleTask process:16:17:44.328098600
scheduling-1--simpleTask process:16:17:51.349931
```

将 fixedDelay = 2000 改成 fixedRate = 2000,再次测试。

```
@Scheduled(initialDelay = 10000,fixedRate = 2000)
public void simpleTask() throws InterruptedException {
    String threadName=Thread.currentThread().getName();
    System.out.println(threadName+"--simpleTask process:" + LocalTime.now());
    Thread.sleep(5000);       //模拟长时间执行,休眠 5000 毫秒
}
```

运行 Springdemo6Application 类，项目启动 10000 毫秒后，控制台输出如下信息。fixedRate＝2000 表示在上一次任务开始执行后间隔 2000 毫秒执行下次任务。而 simpleTask 运行时间为 5000 毫秒，大于 fixedRate，由于任务同步执行，只能等待上一次任务执行完毕再执行下次任务。因此两次任务执行时间为 5000 毫秒，fixedRate 设置的值失效。

```
scheduling-1--simpleTask process:16:29:26.279332100
scheduling-1--simpleTask process:16:29:31.289699600
scheduling-1--simpleTask process:16:29:36.297058100
```

下面在 simpleTask 方法上使用注解@Scheduled(cron = "0/2 * * * * ?")修饰，设置每 2 秒执行一次 simpleTask 方法。

```
@Scheduled(cron = "0/2 * * * *?")
public void simpleTask() {
    String threadName=Thread.currentThread().getName();
    System.out.println(threadName+"--method1 process:" + LocalTime.now());
    Thread.sleep(5000);       //模拟长时间执行,休眠 5000 毫秒
}
```

运行 Springdemo6Application 类，控制台输出如下信息。与 fixedRate 类似，两次任务执行时间为 5000 毫秒。

```
scheduling-1--method1 process:16:38:12.009753200
scheduling-1--method1 process:16:38:17.010725700
scheduling-1--method1 process:16:38:22.023706200
```

上述案例为单任务调度，同步执行场景。如果有 A 和 B 两个方法同时被@scheduled 修饰，两个定时任务同时存在。基于单线程的多任务调度必然存在阻塞的问题，如果 A 方法执行时间过长，B 方法执行会被阻塞直到 A 方法执行完成。同时由于任务同步执行，同一任务的定时执行也会被自身阻塞，即 A 阻塞 A 的下次任务执行，B 阻塞 B 的下次任务执行。

这里修改 ThreadTask 如下：注释掉原有的 simpleTask 方法，新增 method1 和 method2 方法。method1 方法休眠 5000 毫秒。method1 和 method2 方法上都添加@Scheduled (cron="0/5 * * * * ?")，共同执行定时任务。

```
package springdemo6.task;
import org.springframework.scheduling.annotation.Scheduled;
import org.springframework.stereotype.Component;
import java.time.LocalTime;
@Component
public class ThreadTask {
    @Scheduled(cron = "0/5 * * * *?")
    public void method1() throws InterruptedException {
        String threadName=Thread.currentThread().getName();
        System.out.println(threadName+"--method1 process:" + LocalTime.now());
```

```
        Thread.sleep(5000);      //模拟长时间执行,休眠 5000 毫秒
    }
    @Scheduled(cron = "0/5 * * * *?")
    public void method2() {
        String threadName=Thread.currentThread().getName();
        System.out.println(threadName+"--method2 process:" + LocalTime.now());
    }
}
```

再次运行 Springdemo6Application 类,控制台输出如下信息。首先由于 method1 和 method2 方法都使用同一个线程池中的同一个线程 scheduling-1 调度,method1 方法阻塞会造成整个线程阻塞,method2 后续任务执行都受到影响,method2 时间间隔为 10000 毫秒。其次,由于任务同步执行,method1 自身定时任务执行也受到阻塞,method1 时间间隔也为 10000 毫秒。

```
scheduling-1--method2 process:10:47:35.013697300
scheduling-1--method1 process:10:47:35.014711900
scheduling-1--method2 process:10:47:40.026659400
scheduling-1--method1 process:10:47:45.011868700
scheduling-1--method2 process:10:47:50.026659300
scheduling-1--method1 process:10:47:55.011689500
scheduling-1--method2 process:10:48:00.021441500
```

通过上述案例,可以看到 SpringTask 默认的单线程任务调度和同步执行机制在执行多任务时存在安全隐患,如果同时有多个任务发生不同程度的阻塞,会造成连锁反应,导致所有任务执行错乱。因此,在多任务场景中,需对 SpringTask 任务机制进行优化。

6.2.3　Spring Task 基于多个定时任务编程实现

在多任务场景中,对 SpringTask 任务机制进行优化,可以从任务调度和任务执行两方面入手。

1. 从任务调度方面进行优化

有如下两种实现方案。

方案 1:增加原任务调度线程池 ScheduledThreadPoolExecutor 的线程数量,使之大于 1。

方案 2:自定义一个线程池,完全替换原有的单线程池。

方案 1 的实现方法如下。在 config 目录下新建 ScheduledConfig 配置类,内部覆写 configureTasks 方法,增加原有任务调度线程池 ThreadPoolTaskScheduler 的线程数量。

【ScheduledConfig.java】

```
package springdemo6.config;
import org.springframework.context.annotation.Configuration;
import org.springframework.scheduling.annotation.SchedulingConfigurer;
import org.springframework.scheduling.config.ScheduledTaskRegistrar;
import java.util.concurrent.Executors;
@Configuration
```

```java
public class ScheduledConfig implements SchedulingConfigurer {
    @Override
    public void configureTasks(ScheduledTaskRegistrar taskRegistrar) {
        //增加原有线程池的线程数量为 10
        taskRegistrar.setScheduler(Executors.newScheduledThreadPool(10));
    }
}
```

运行 Springdemo6Application,控制台输出结果如下。可以看到 method1 的阻塞没有影响 method2 的执行,method2 时间间隔为 5000 毫秒,两个定时任务之间不再互相干扰。但是 method1 自身的定时任务仍然以同步方式运行,每次执行都需等待上次执行结束。

```
pool-1-thread-2--method2 process:14:24:35.003667500
pool-1-thread-1--method1 process:14:24:35.003667500
pool-1-thread-2--method2 process:14:24:40.009586700
pool-1-thread-3--method1 process:14:24:45.007781500
pool-1-thread-1--method1 process:14:24:45.007781500
pool-1-thread-2--method2 process:14:24:50.007153200
pool-1-thread-4--method1 process:14:24:55.002298500
```

方案 2 的实现方法如下。自定义一个线程池,这里使用 ThreadPoolTaskScheduler 线程池,该线程池底层使用 JDK 的 ScheduledThreadPoolExecutor 线程池实现。在 config 目录下新建配置类 MySchedulerPool,内部自定义一个名为 MySchedulerPool 的线程池。

【MySchedulerPool.java】

```java
package springdemo6.config;
import org.springframework.context.annotation.Bean;
import org.springframework.context.annotation.Configuration;
import org.springframework.scheduling.TaskScheduler;
import org.springframework.scheduling.concurrent.ThreadPoolTaskScheduler;
@Configuration
public class MySchedulerPool {
    //自定义一个线程池 myThreadPoolTaskScheduler
    @Bean(name="MySchedulerPool")
    public TaskScheduler myThreadPoolTaskScheduler() {
        ThreadPoolTaskScheduler taskScheduler = new ThreadPoolTaskScheduler();
        taskScheduler.setPoolSize(10);
        taskScheduler.setThreadNamePrefix("myThreadPool-");
        return taskScheduler;
    }
}
```

修改 ScheduledConfig 配置类,在内部注入自定义线程池对象,设置 ScheduledTaskRegistrar 使用自定义线程池做任务调度。

```java
package springdemo6.config;
import org.springframework.beans.factory.annotation.Autowired;
import org.springframework.beans.factory.annotation.Qualifier;
```

```java
import org.springframework.context.annotation.Configuration;
import org.springframework.scheduling.TaskScheduler;
import org.springframework.scheduling.annotation.SchedulingConfigurer;
import org.springframework.scheduling.config.ScheduledTaskRegistrar;
import java.util.concurrent.Executors;
@Configuration
public class ScheduledConfig implements SchedulingConfigurer {
    @Autowired
    @Qualifier(value = "MySchedulerPool")
    private TaskScheduler taskScheduler;
    @Override
    public void configureTasks(ScheduledTaskRegistrar taskRegistrar) {
        //增加原有线程池的线程数量为 10
        taskRegistrar.setScheduler(Executors.newScheduledThreadPool(10));
        //注册自定义线程池进行任务调度
        taskRegistrar.setTaskScheduler(taskScheduler);
    }
}
```

运行 Springdemo6Application，控制台输出结果和方案 1 一样，解决了 method1 和 method2 任务之间的干扰，但是没有解决 method1 自身的任务同步执行问题。

```
myThreadPool-1--method1 process:15:20:40.002318600
myThreadPool-2--method2 process:15:20:40.002318600
myThreadPool-2--method2 process:15:20:45.012236900
myThreadPool-1--method2 process:15:20:50.013743200
myThreadPool-3--method1 process:15:20:50.013743200
myThreadPool-2--method2 process:15:20:55.004020500
myThreadPool-4--method2 process:15:21:00.003908700
myThreadPool-5--method1 process:15:21:00.003908700
```

2. 从任务执行方面优化

从任务执行方面优化，即不改变任务调度的默认线程池，利用线程池将任务同步执行方式改为异步执行，实现异步非阻塞运行。最常见的实现方式就是在定时任务方法上使用@Async 注解配合自定义线程池实现。@Async 注解修饰的方法在每次执行时，都从线程池中取出新的线程异步执行，消除了线程阻塞。这里使用 ThreadPoolTaskExecutor 线程池做自定义线程池，这也是 Spring 推荐使用的线程池。ThreadPoolTaskExecutor 本质上是对 JDK 中 ThreadPoolExecutor 的封装。使用时可直接在 application.yaml 中配置 ThreadPoolTaskExecutor 线程池的相关参数，ThreadPoolTaskExecutor 相关配置如下。

```
spring:
  task:
    execution:
      pool:
        allow-core-thread-timeout: true    //允许核心线程超时,动态增加和缩小线程池
        max-size: 100                      //线程池最大线程数
        core-size: 8                       //线程池核心线程数,默认为 8
```

```yaml
      keep-alive: 60s                    //线程终止前允许保持空闲的时间
      queue-capacity: 100                //线程池队列容量大小
    shutdown:
      await-termination: false           //是否等待所有任务完成后才关闭应用
      await-termination-period: 60s      //等待所有任务完成的最大时间
    thread-name-prefix: NewThread        //用于新创建线程名称的前缀
```

在 application.yaml 中配置上述 ThreadPoolTaskExecutor 线程池后,为避免干扰,注释 MySchedulerPool 和 ScheduledConfig 类上的@Configuration 注解,使之都不生效。然后在 ThreadTask 的 method1 和 method2 方法上都添加@Async 注解,以异步方式执行。

```java
@Scheduled(cron = "0/5 * * * * ?")
@Async
public void method1() throws InterruptedException {
    String threadName=Thread.currentThread().getName();
    System.out.println(threadName+"--method1 process:" + LocalTime.now());
    Thread.sleep(5000);                  //模拟长时间执行,休眠 5000 毫秒
}
@Scheduled(cron = "0/5 * * * * ?")
@Async
public void method2() {
    String threadName=Thread.currentThread().getName();
    System.out.println(threadName+"--method2 process:" + LocalTime.now());
}
```

运行 Springdemo6Application 类,控制台输出如下信息,可以看到 method1 和 method2 每次执行时都从线程池中取出了不同的线程执行,互不干扰。虽然 method1 延迟阻塞,但是 method1 和 method2 后续任务执行都没有受到影响,method1 和 method2 的执行时间间隔都为 5 秒。完美解决了线程阻塞问题。

```
NewThread1--method2 process:14:03:15.016837200
NewThread2--method1 process:14:03:15.016837200
NewThread3--method2 process:14:03:20.008084900
NewThread4--method1 process:14:03:20.008084900
NewThread5--method1 process:14:03:25.002598400
NewThread6--method2 process:14:03:25.003595800
NewThread7--method2 process:14:03:30.002112200
NewThread8--method1 process:14:03:30.002112200
```

6.2.4 Spring Task 动态定时任务编程实现

在前面的介绍中,我们设置的定时任务执行时间都是固定的。如果要改变定时任务执行时间,只能手动在代码中修改再重启项目,这非常不方便。如果应用场景中需要对任务执行时间做即时修改,修改完毕不需重启项目即时生效。使用@Scheduled 注解就不能满足需求。这里可结合 ThreadPoolTaskSchedulerschedule 线程池和 ScheduledFuture 实现功能,首先需手动结束之前的定时任务,随后更新时间重新开启新的定时任务。具体实现时可先利用

ThreadPoolTaskSchedulerschedule 任务调度线程池的 schedule 方法开启定时任务,开启的定时任务用 ScheduledFuture 保存。在动态修改定时任务执行时间后先调用 ScheduledFuture 的 cancel 方法结束旧的定时任务,再以修改后的新时间开启新的定时任务。

具体的编程实现如下。在 config 目录下新建 MyThreadPoolTaskScheduler 类,内部定义一个线程池 myThreadPoolTaskSchedulerPool,以便后续定时任务由多线程执行。如不新建该类,后续定时任务由单线程执行。

【MyThreadPoolTaskScheduler.java】

```
package springdemo6.config;
import org.springframework.context.annotation.Bean;
import org.springframework.context.annotation.Configuration;
import org.springframework.scheduling.concurrent.ThreadPoolTaskScheduler;
@Configuration
public class MyThreadPoolTaskScheduler {
    //自定义一个线程池
    @Bean(name="myThreadPoolTaskSchedulerPool")
    public ThreadPoolTaskScheduler MyThreadPoolTaskSchedulerPool() {
        ThreadPoolTaskScheduler taskScheduler = new ThreadPoolTaskScheduler();
        taskScheduler.setPoolSize(10);
        taskScheduler.setThreadNamePrefix("DiyMyThreadPool-");
        return taskScheduler;
    }
}
```

在 task 目录下新建 DynamicTask 类,内部注入的 myThreadPoolTaskSchedulerPool 线程池用于执行定时任务。DynamicTask 类内部定义的 start 方法开始执行定时任务,定义的 setCron 方法动态修改定时任务的执行时间。

【DynamicTask.java】

```
package springdemo6.task;
import org.springframework.beans.factory.annotation.Autowired;
import org.springframework.context.annotation.Configuration;
import org.springframework.scheduling.Trigger;
import org.springframework.scheduling.TriggerContext;
import org.springframework.scheduling.annotation.EnableScheduling;
import org.springframework.scheduling.concurrent.ThreadPoolTaskScheduler;
import org.springframework.scheduling.support.CronTrigger;
import java.time.Instant;
import java.time.LocalTime;
import java.util.concurrent.ScheduledFuture;
@Configuration
@EnableScheduling
public class DynamicTask{
    private String cron="0/2 * * * * ?";
    //注入线程池 ThreadPoolTaskScheduler。如没有配置线程池,则以单线程执行任务
    @Autowired(required = false)
    ThreadPoolTaskScheduler scheduler;
```

```java
    private ScheduledFuture<?> future;
    public String getCron() {
        return cron;
    }
    public void setCron(String cron) {
        this.cron = cron;
        System.out.println("设置任务定时时间为"+
                cron.split(" ")[0].split("/")[1]+"秒,当前时间为"+ LocalTime.now());
        if(future!=null){
            future.cancel(true);
        }
        start();
    }
    //开始执行定时任务
    private void start() {
        //定义一个定时任务 runnable
        Runnable runnable=new Runnable() {
            @Override
            public void run() {
                String threadname=Thread.currentThread().getName();
                System.out.println(threadname+"--DynamicTask:" + LocalTime.now());
            }
        };
        //定义一个触发器触发定时任务
        Trigger trigger=new Trigger() {
            @Override
            public Instant nextExecution(TriggerContext triggerContext) {
                CronTrigger cronTrigger = new CronTrigger(cron);
                Instant nextExec = cronTrigger.nextExecution(triggerContext);
                return nextExec;
            }
        };
        //利用线程池的 ThreadPoolTaskScheduler 的 scheduler 方法执行定时任务
        future = scheduler.schedule(runnable,trigger);
    }
}
```

在 task 目录下新建 StartDynamicTask 类,StartDynamicTask 类继承 ApplicationListener 接口,在 Spring Boot 容器加载完毕,马上调用 DynamicTask 的 start 方法执行定时任务。StartDynamicTask 类内部创建一个线程 runnable,每隔 20000 毫秒修改一次定时任务执行时间,时间从 list 集合中随机选取。

【StartDynamicTask.java】

```java
package springdemo6.task;
import org.springframework.beans.factory.annotation.Autowired;
import org.springframework.context.ApplicationEvent;
import org.springframework.context.ApplicationListener;
import org.springframework.stereotype.Component;
```

```java
import java.util.List;
import java.util.Random;
import java.util.stream.Stream;
@Component
public class StartDynamicTask implements ApplicationListener {
    @Autowired
    private DynamicTask dynamicTask;
    @Override
    public void onApplicationEvent(ApplicationEvent event) {
        //开始执行定时任务
        dynamicTask.setCron("0/2 * * * * ?");
        List<Integer> list= Stream.of(1,2,4,6).toList();
        //创建线程,内部每隔 20000 毫秒修改一次时间
        Runnable runnable=new Runnable() {
            @Override
            public void run() {
                while(true){
                    try{
                        int time= new Random().nextInt(list.size());
                        String cron="0/"+list.get(time)+" * * * * ?";
                        dynamicTask.setCron(cron);
                        Thread.sleep(20000);
                    }catch(InterruptedException e){
                        e.printStackTrace();
                    }
                }
            }
        };
        runnable.run();
    }
}
```

运行 Springdemo6Application,控制台输出结果如下。可以看到任务定时时间从 6000 毫秒修改为 4000 毫秒时,DynamicTask 任务执行时间间隔也同步动态修改了。

```
设置任务定时时间为 6000 毫秒,当前时间为 21:02:05.784852900
DiyMyThreadPool-3--DynamicTask:21:02:06.000810300
DiyMyThreadPool-4--DynamicTask:21:02:12.010695100
DiyMyThreadPool-1--DynamicTask:21:02:18.012479
DiyMyThreadPool-5--DynamicTask:21:02:24.007131200
设置任务定时时间为 4000 毫秒,当前时间为 21:02:25.792345400
DiyMyThreadPool-2--DynamicTask:21:02:28.005067100
DiyMyThreadPool-3--DynamicTask:21:02:32.009772500
DiyMyThreadPool-3--DynamicTask:21:02:36.014109400
DiyMyThreadPool-3--DynamicTask:21:02:40.002055400
DiyMyThreadPool-3--DynamicTask:21:02:44.004187700
设置任务定时时间为 1000 毫秒,当前时间为 21:02:45.795219700
DiyMyThreadPool-9--DynamicTask:21:02:46.010872400
DiyMyThreadPool-9--DynamicTask:21:02:47.014690100
DiyMyThreadPool-5--DynamicTask:21:02:48.002559700
```

任务 6.3 综合案例：利用 Spring Task 实现定时闹钟

基于上述知识，本任务以一个案例演示 Spring Task 的综合应用，案例以实现定时闹钟为主要功能，编程中需要用到多线程任务调度的相关知识，以及 Spring Task 动态设置定时任务和 Mybatis 操作数据库的相关知识。

6.3.1 案例任务

基于 SpringDemo6 项目实现类似于苹果手机上的定时闹钟功能，闹钟设置可以输入如下三个参数：星期、具体时分秒时间、闹钟是否重复执行。任务时间执行逻辑采用 cron 表达式设置。如果闹钟重复执行，利用 cron 表达式设置定时任务；如果闹钟不重复执行，利用 cron 表达式设置固定时间点任务。如输入参数为星期一、6:00:00、重复执行，则闹钟在每周一 6:00:00 时刻自动执行。如果输入参数为星期一、6:00:00、不重复执行，则闹钟在下个周一 6:00:00 时刻只自动执行一次。在定时任务执行过程中，可以随时动态添加多个闹钟加入任务计划。添加的闹钟要存入 MySQL 数据库中，以便每次服务启动执行未过期的任务。

6.3.2 案例分析

该任务将项目 5 中数据库操作和本项目的定时任务相关知识相结合，编程中需要根据输入的三个参数（星期、具体时分秒时间、闹钟是否重复执行）编写 cron 表达式，控制每个任务执行的时间逻辑。同时利用多线程实现多个定时任务计划的动态新增和调度执行，且任务计划需要持久化存储。该任务基于 SpringDemo6 项目实现，任务具体的实现思路如下。

(1) 新建数据库和数据表，添加表字段，表字段为闹钟 id(clock_id)、闹钟星期(clock_week)、闹钟时间(clock_time)、是否重复(clock_isrepeat)、是否过期(clock_islose)。

(2) 在项目中添加 Mybatis 相关依赖并在配置文件 application.yaml 中添加数据源相关配置。

(3) 编写 Clock 实体类，并利用 Mybatis 编写闹钟任务的增、改和查操作。

(4) 编写一个闹钟任务执行线程池，以便多线程执行不同闹钟任务。

(5) 编写闹钟任务执行类，内部定义一个任务执行线程和一个触发器用于定时执行任务。如果任务不重复执行，任务执行一次后，需修改数据库中该任务的字段 clock_islose 为 true。

(6) 编写闹钟任务管理类，内部利用缓存存储任务信息。如果有新任务添加，则利用闹钟任务执行类封装任务，将该任务执行类放进线程池执行。

(7) 编写闹钟任务启动类，在项目开始自动读取数据库中所有未过期(clock_islose 为 false)的闹钟任务，依次利用任务执行类封装，并添加进线程池执行。

(8) 编写测试类，不断往里添加新的闹钟任务，进行测试。

6.3.3 任务实施

由于篇幅有限,任务实施的具体步骤采用二维码形式展示。

综合案例实施步骤

小 结

本项目中详细介绍了基于 Spring Boot 项目的定时任务编程实现,包括 Cron 表达式的使用和 SpringTask 定时任务编程。重点介绍基于多线程的 SpringTask 静态和动态定时任务调度编程。最后以一个综合案例——Spring Task 实现定时闹钟,演示如何在 Spring Boot 项目中综合运用 Mybatis 并结合 Spring Task 在多线程场景下实现闹钟任务的自由调度和存储,使读者进一步掌握 Spring Boot 的数据读写编程。

课后练习:定时清除过期闹钟任务

基于本项目的任务 6.3 添加功能,在每晚上 12 点定时清除 ConcurrentHashMap 缓存中和数据库中的过期闹钟任务。

项目 7 Spring Boot 消息队列

在前面的章节中,我们利用 Spring Boot 构建了很多的微服务应用。如果这些微服务应用之间要进行消息通信,由于各微服务运行并不是同步的,采用同步模式会产生大量等待时间,影响性能,因此大多采用异步处理模式。在异步处理模式中,假设服务 A 和服务 B 进行通信,为保证服务 A 发送的信息不丢失,需要将服务 A 发送的消息先有序地以缓存形式保存在一个中间件中,以便服务 B 在空闲的时间能够有序地处理消息,反之亦然。这个中间件就是一个消息队列。消息队列的具体实现很多,如 RabbitMQ、ActiveMQ、RocketMQ 和 Kafka 等。本项目以消息队列框架 Kafka 为代表,介绍如何在 Spring Boot 项目中使用消息队列进行编程实践。

任务 7.1 初识消息队列

在使用消息队列编程之前,先要了解消息队列。本任务将介绍消息队列的相关理论知识,包括消息队列的基本概念,为什么要使用消息队列以及常用的主流消息队列框架,以便后续在编程中应用。

7.1.1 消息队列简介

消息队列是在消息的传输过程中保存消息的容器。在消息的通信过程中,除了传统的请求响应模式外,还有生产消费模式。消息队列的实现是基于生产消费模式的消息通信,如图 7-1 所示。在生产消费模式中,消息发送方为生产者,消息接收方为消费者。由于生产者和消费者各自采用异步模式处理消息,就需要一个中间件来对双方发送的消息进行顺序存储,其中每条消息通过唯一 id 进行区分。生产者可以随时往中间件中发送消息,不必关注消费者什么时候接收处理。当消费者需要取出消息进行处理时,也可以随时从中间件中取出消息。消息既不会丢失,又能最大限度地提升消息处理速度。消息队列是开发大型分布式应用系统中必备的组件,使用消息队列能够进一步提升异步模式的消息处理速度,实现流量削峰和降低系统耦合。

图 7-1 消息队列

7.1.2 常用的消息队列中间件

消息队列中间件的具体实现方式有多种，主要分为以下三种实现方式。

1. 基于 JMS 协议

JMS 即 Java 消息服务（Java message service）协议，用于在多个服务之间进行异步通信。当多个服务之间需要进行通信时，可以使用 JMS 作为中间件进行消息的存储和转发，降低服务之间的耦合。为方便使用 JMS，Spring 中提供了 spring-jms 组件对 Java 中的 JMS 相关 API 进行进一步封装。JMS 提供点对点和发布订阅两种消息发送方式。点对点方式下，消息生产者将消息将发送到一个队列，该队列的消息只能被一个消费者消费。发布订阅方式下，生产者发送的消息带有特定的主题，多个相同主题的消费者可以共同消费同一主题数据。基于 JMS 实现的消息队列中间件有 ActiveMQ 和 HornetMQ 等。

2. 基于 AMQP

高级消息队列协议（advanced message queuing protocol，AMQP），内部也兼容 JMS 协议。AMQP 是一个链接协议，定义的是网络交换的数据格式，而不关心具体的实现语言。与 JMS 相比，AMQP 具备跨平台和跨语言的天然优势。如可以用 Java 语言编写生产者，用 Python 语言编写消费者，一样能够实现消息的传递，这点和 HTTP 很像。AMQP 利用交换机、捆绑和队列等元素实现消息传递。生产者将消息发送给交换机，由捆绑决定交换机的消息发送到哪个队列，每个队列指定特定的路由键。消费者监听特定路由键的队列，如有消息到达，直接从队列中消费消息。基于 AMQP 实现的消息队列中间件有 RabbitMQ、RocketMQ 等。

3. 基于自定义协议

Kafka 是一款开源且基于分布式架构的发布订阅消息系统，单节点支持百万级数据吞吐量。因其高吞吐量、高扩展性的优点，经常被用作消息队列中间件。Kafka 的底层采用自定义的 TCP 二进制协议进行消息通信。Kafka 设计的初衷是为了满足大数据领域实时日志处理，但也能够运用在其他景下实现消息队列。Kafka 中的每条消息都有一个主题，可基于特定主题创建生产者和消费者对象，消费者只能消费本主题的数据。

在实际应用中，如果消息并发量很大，达到百万级可使用 Kafka，一般情况下大多用 RabbitMQ 和 RocketMQ。后续内容以 Kafka 为代表，介绍如何在 Spring Boot 项目中使用消息队列进行编程。

任务 7.2　基于 Kafka 的消息队列编程

Kafka 是一款主流、高性能的开源分布式发布订阅消息系统，常用来作为消息中间件存储和转发消息。本任务将重点介绍如何在 Spring Boot 项目中利用 Kafka 实现消息传递。

7.2.1　Kafka 简介

Kafka 是一款使用 Scala 语言编写的分布式发布订阅消息系统，主要运用于大数据领域的实时数据采集和处理。由于 Kafka 的高吞吐量和低延迟特性，一般用于高并发场景下的消息传递。Kafka 基于主题机制收发数据。Kafka 整体架构如图 7-2 所示。

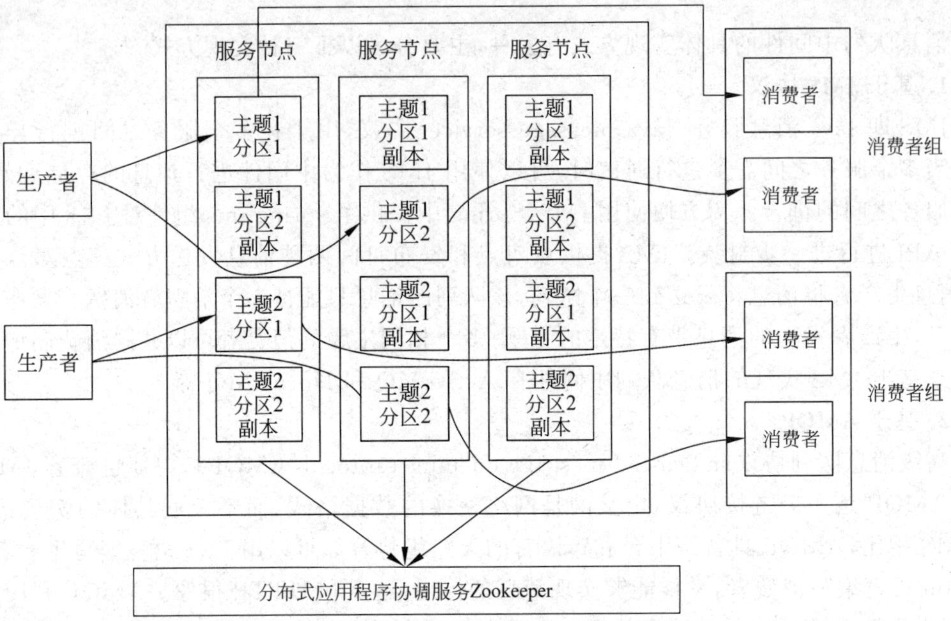

图 7-2 Kafka 架构图

图 7-2 中的核心概念说明如下。

(1) 生产者(producer)：发布消息的对象。

(2) 服务节点(broker)：Kafka 集群服务，一个 Kafka 集群服务由多个 broker 组成。

(3) 主题(topic)：Kafka 基于主题分类消息，每类消息有一个唯一的主题。一个 broker 可以有多个主题。

(4) 分区(partition)：如果一个主题消息量过大，主题可以设置多个分区存储消息，不同的分区消息存放在不同的节点上，实现负载均衡。分区中的每个消息都有一个唯一且连续的序号，用于标识一条消息。分区内通过消息偏移量(offset)记录下一条将要发送给消费者的消息序号，消息偏移量默认初始值从 0 开始，分区内每发送一条消息，消息偏移量加 1。

(5) 副本(replication)：每一个分区都有多个副本，当分区故障时，可以使用副本内数据快速替代，保证数据不丢失，这也是分布式架构常用的容错机制。

(6) 消费者(consumer)：消费消息的对象。消费者可通过多种灵活的消费方式消费消息，可从主题、分区甚至指定的偏移量消费消息。消费者内部通过消费偏移量(consumer offset)标识消费消息的位置，消费偏移量默认初识值从 0 开始，每消费一条消息，消费偏移量加 1。在新版 Kafka 中以消费者组为单位消费消息，一个消费者组默认包含一个消费者，也可设置多个消费者。同一主题、同一分区的数据只能被消费者组中的某一个消费者消费，同一个消费者组的多个消费者可以消费同一主题、不同分区的数据，这样能够提高数据消费速度，提升吞吐量。

(7) 分布式应用程序协调服务(Zookeeper)：Kafka 没有自己的分布式协调服务组件，Kafka 集群默认使用 Zookeeper 来管理集群的元数据信息，如集群信息、主题信息、分区信息、消息偏移量等,保证服务的一致性。

Kafka 相关概念

7.2.2 Kafka 安装和配置

下面介绍如何在 Windows 系统下安装 Kafka 并配置,具体步骤请扫描二维码观看。

Kafka 安装和配置

7.2.3 Spring Boot 引入 Kafka

在 Spring Boot 项目中使用 Kafka 只需在项目 pom.xml 文件中引入如下 Kafka 依赖即可。Spring Boot 会对 Kafka 进行自动配置。

```xml
<dependency>
    <groupId>org.springframework.kafka</groupId>
    <artifactId>spring-kafka</artifactId>
</dependency>
```

如需修改 Kafka 默认配置,可以在项目的 application.yaml 配置文件中修改。Kafka 常用配置如下。

```yaml
kafka:
  bootstrap-servers: localhost:9092        //Kafka 服务器地址
  consumer:
    group-id: kafkaconsumergroup           //消费者组 id
    //消费者组关键字的序列化类,默认序列化字符串
    key-deserializer: org.apache.kafka.common.serialization.StringDeserializer
    //消费者组值的序列化类,默认序列化字符串
    value-deserializer: org.apache.kafka.common.serialization.StringDeserializer
  producer:
    //生产者关键字的序列化类,默认序列化字符串
    key-serializer: org.apache.kafka.common.serialization.StringSerializer
    //生产者值的序列化类,默认序列化字符串
    value-serializer: org.apache.kafka.common.serialization.StringSerializer
```

这里先创建一个 SpringDemo7 项目,在项目的 pom.xml 文件中引入上述 Kafka 依赖,

Kafka 相关配置均使用默认配置,后续编程将在该项目中进行。

7.2.4 Spring Boot 基于 Kafka 的编程实现

为方便 Kafka 的编程,Spring 提供 KafkaTemplate 模板对象用于消息的发送和接收,KafkaTemplate 内部定义了许多数据收发的方法,使用时只需在 Java 类中注入 RabbitTemplate 对象即可使用。Kafka 的消息传递是基于主题实现的,生产者基于主题发送消息,消费者支持从主题、分区和偏移量三个维度消费消息。下面结合不同消费模式演示 Kafka 编程实现。这里假设已经分别启动了 Zookeeper 和 Kafka 服务。

1. 单线程消费一个主题

在 SpringDemo7 项目的 springdemo7 目录下新建 producers 目录,在 producers 目录下新建 KafkaProducer 类,类中注入 KafkaTemplate 对象,并调用 send 方法向 kafkatopic 主题的 0 号分区发送 2 条消息。send 方法有很多输入参数,最简单的一种是只传入主题和消息内容。为方便介绍,这里 KafkaProducer 类中的 send 方法传入了所有参数。

【KafkaProducer.java】

```java
package springdemo7.producers;
import org.springframework.beans.factory.annotation.Autowired;
import org.springframework.kafka.core.KafkaTemplate;
import org.springframework.stereotype.Component;
@Component
public class KafkaProducer {
    @Autowired(required = false)
    private KafkaTemplate kafkaTemplate;
    public void send() {
        String msg1="Send Kafka Message1";
        String msg2="Send Kafka Message2";
        long now=System.currentTimeMillis();
        //主题为 kafkatopic,分区号为 0,时间为当前时间 now,key 为 kafkatopic,消息内容
        //为 msg1 和 msg2
        kafkaTemplate.send("kafkatopic",0,now,"kafkatopic",msg1);
        kafkaTemplate.send("kafkatopic",0,now,"kafkatopic",msg2);
    }
}
```

Kafka 消费数据有两种模式,分别是推模式和拉模式。推模式由 Kafka 主动推送消息给消费者,消费者使用@KafkaListener 注解监听特定主题、分区或偏移量获取消息,实时性好,消费者能及时得到最新的消息。拉模式由消费者主动调用 KafkaTemplate 的 receive 方法消费消息,实时性差,消费者难以获取实时消息。实际应用中一般使用推模式消费数据。推模式最简单的实现方法就是在直接在消费数据的方法上使用@KafkaListener 注解。

下面在 springdemo7 目录下新建 consumers 目录,在 consumers 目录下新建 KafkaConsumer 类,类中定义 receiveMsg 方法消费消息。在 receiveMsg 方法上添加注解@KafkaListener(topics = {"kafkatopic"}, groupId = "kafkaconsumergroup"),开启一个线程实时监听 kafkatopic 主题,并指定消费者组为 kafkaconsumergroup。

【KafkaConsumer.java】
```
package springdemo7.consumers;
import org.apache.kafka.clients.consumer.ConsumerRecord;
import org.springframework.kafka.annotation.KafkaListener;
import org.springframework.stereotype.Component;
@Component
public class KafkaConsumer {
    @KafkaListener(topics = {"kafkatopic"},groupId = "kafkaconsumergroup")
    public void receiveMsg(ConsumerRecord consumerRecord) {
        System.out.println("=========主题:"+consumerRecord.topic()+"\n"+
                "分区:"+consumerRecord.partition()+"\n"+
                "消费偏移量:"+consumerRecord.offset()+"\n"+
                "key:"+consumerRecord.key()+"\n"+
                "消息内容:"+consumerRecord.value()+"\n"+
                "创建消息时间戳:"+consumerRecord.timestamp()+"\n"+
                "线程:"+Thread.currentThread().getName());
    }
}
```

在 springdemo7 目录下新建测试类 KafkaTest,该类内部实现 ApplicationListener 接口并监听 ApplicationReadyEvent 事件,在项目启动后延迟 5000 毫秒向 Kafka 发送信息。

【KafkaTest.java】
```
package springdemo7;
import org.springframework.beans.factory.annotation.Autowired;
import org.springframework.boot.context.event.ApplicationReadyEvent;
import org.springframework.context.ApplicationListener;
import org.springframework.stereotype.Component;
import springdemo7.producers.KafkaProducer;
@Component
public class KafkaTest implements
        ApplicationListener<ApplicationReadyEvent> {
    @Autowired
    private KafkaProducer kafkaProducer;
    @Override
    public void onApplicationEvent(ApplicationReadyEvent event) {
        try{
            Thread.sleep(5000);    //延迟 5000 毫秒发送消息
            kafkaProducer.send();
        }catch(InterruptedException e){
            e.printStackTrace();
        }
    }
}
```

启动 Springdemo7Application,等待 5000 毫秒后在控制台可以看到使用一个线程接收到 kafkatopic 主题的 0 号分区的两条消息,分别是 Send Kafka Message1 和 Send Kafka Message2,消费者组 kafkaconsumergroup 对 kafkatopic 主题第一条消息消费,消费偏移量为 0;第二条消息在第一条消息基础上消费,偏移量加 1。

```
=========主题:kafkatopic
分区:0
消费偏移量:0
key:kafkatopic
消息内容:Send Kafka Message1
创建消息时间戳:1682341231854
线程:org.springframework.kafka.KafkaListenerEndpointContainer#0-0-C-1
=========主题:kafkatopic
分区:0
消费偏移量:1
key:kafkatopic
消息内容:Send Kafka Message2
创建消息时间戳:1682341231854
线程:org.springframework.kafka.KafkaListenerEndpointContainer#0-0-C-1
```

2. 单线程消费多个主题

除了消费一个主题外,Kafka 消费者还可以同时消费多个主题数据。这里在 KafkaProducer 类中添加 sendTwoTopic 方法,内部同时向 kafkatopic1 主题和 kafkatopic2 主题发送消息。

```
//向两个不同主题发数据
public void sendTwoTopic() {
    String msg1="Send Kafka Message1";
    String msg2="Send Kafka Message2";
    long now=System.currentTimeMillis();
    //主题为 kafkatopic1,分区号为 0,时间为当前时间 now,key 为 kafkatopic1,消息内容
        //为 msg1
    kafkaTemplate.send("kafkatopic1",0,now,"kafkatopic1",msg1);
    //主题为 kafkatopic2,分区号为 0,时间为当前时间 now,key 为 kafkatopic3,消息内容
    //为 msg2
    kafkaTemplate.send("kafkatopic2",0,now,"kafkatopic2",msg2);
}
```

在 KafkaConsumer 类中修改原有的 @KafkaListener 注解内容如下:同时监听 kafkatopic1 主题和 kafkatopic2 主题。

```
@KafkaListener(topics = {"kafkatopic1","kafkatopic2"},groupId = "kafkaconsumergroup")
```

将 TestKafka 类中 kafkaProducer 调用 send 方法改为调用 sendTwoTopic 方法。

```
kafkaProducer.sendTwoTopic();
```

启动 Springdemo7Application,等待 5000 毫秒后,在控制台可以看到使用一个线程接收到 Kafkatopic1 和 Kafkatopic2 为主题各 1 条数据。由于 Kafkatopic1 和 Kafkatopic2 为新建主题, 消费者组 kafkaconsumergroup 对 Kafkatopic1 主题的消费偏移量为 0。对 Kafkatopic2 主题的消费偏移量为 0。

```
=========主题:kafkatopic1
分区:0
消费偏移量:0
key:kafkatopic1
消息内容:Send Kafka Message1
创建消息时间戳:1682341787226
线程:org.springframework.kafka.KafkaListenerEndpointContainer#0-0-C-1
=========主题:kafkatopic2
分区:0
消费偏移量:0
key:kafkatopic2
消息内容:Send Kafka Message2
创建消息时间戳:1682341787226
线程:org.springframework.kafka.KafkaListenerEndpointContainer#0-0-C-1
```

3. 多线程消费多个主题不同分区

除了消费多个主题外,Kafka 消费者还可以同时消费多个主题的不同分区消息,同时为提升消息消费速度,还可以设置多线程并发消费消息。Kafka 默认创建只有一个分区的主题,这里需通过自定义方式来创建主题,并设置主题分区数。在 springdemo7 目录下新建 config 目录,在 config 目录下新建 Kafka 配置类 KafkaConfig,内部创建 kafkatopic1 和 kafkatopic2 两个主题,每个主题设置分区数为 2,副本数为 2。

【KafkaConfig.java】

```java
package springdemo7.config;
import org.apache.kafka.clients.admin.NewTopic;
import org.springframework.context.annotation.Bean;
import org.springframework.context.annotation.Configuration;
@Configuration
public class KafkaConfig {
    //创建 kafkatopic1 主题,设置 2 个分区的 2 个副本
    @Bean
    public NewTopic initialBHeartBeatMessageTopic() {
        return new NewTopic("kafkatopic1",2, (short) 2);
    }
    //创建 kafkatopic2 主题,设置 2 个分区的 2 个副本
    @Bean
    public NewTopic initialBHeartBeatMessageTopic2() {
        return new NewTopic("kafkatopic2",2, (short) 2);
    }
}
```

在 KafkaProducer 类中添加 sendTopicTwoPartition 方法,内部同时向 kafkatopic1 主题和 kafkatopic2 主题的 0 号分区和 1 号分区发送 2 条消息。

```java
//向不同主题不同分区发数据
public void sendTopicTwoPartition() {
    String msg1="Send Kafka Message1";
```

```
        String msg2="Send Kafka Message2";
        long now=System.currentTimeMillis();
        //向主题为 kafkatopic1 的 0 号分区发消息
        kafkaTemplate.send("kafkatopic1",0,now,"kafkatopic",msg1);
        //向主题为 kafkatopic1 的 1 号分区发消息
        kafkaTemplate.send("kafkatopic1",1,now,"kafkatopic",msg1);
        //向主题为 kafkatopic2 的 0 号分区发消息
        kafkaTemplate.send("kafkatopic2",0,now,"kafkatopic",msg2);
        //向主题为 kafkatopic2 的 1 号分区发消息
        kafkaTemplate.send("kafkatopic2",1,now,"kafkatopic",msg2);
}
```

在 KafkaConsumer 中修改原有 @KafkaListener 注解的内容如下: 同时监听 kafkatopic1 主题和 kafkatopic2 主题的 0 号和 1 号分区,这里设置 concurrency = "2",创建两个消费者,每个消费者开启一个线程并发消费消息。实际使用时,concurrency 值一般等于主题的分区数,一个消费者消费一个分区。若 concurrency 小于分区数,则一个消费者消费多个分区,消费不均。若 concurrency 大于分区数,则有的消费者没有分区消费,浪费资源。

```
@KafkaListener(topicPartitions = {
    @TopicPartition(topic = "kafkatopic1", partitions = {"0", "1"}),
    @TopicPartition(topic = "kafkatopic2", partitions = {"0", "1"}),
},groupId = "kafkaconsumergroup",concurrency = "2")
```

将 KafkaTest 类中的 kafkaProducer 调用 sendTwoTopic 方法改为调用 sendTopicTwoPartition 方法。

```
kafkaProducer.sendTopicTwoPartition();
```

启动 Springdemo7Application,等待 5000 毫秒后,在控制台可以看到使用不同的两个线程分别接收 Kafkatopic1 和 Kafkatopic2 主题的各 2 条消息。由于消费偏移量是保存在 Zookeeper 中的,即使消费者组重启,仍然能够按顺序继续消费数据。对于消费者组 kafkaconsumergroup 来说,由于 Kafkatopic1 和 Kafkatopic2 主题的 0 号分区在上个案例中消费了一次消息,因此这次消费偏移量为 1。而 Kafkatopic1 和 Kafkatopic2 主题的 1 号分区为新建分区,消费者组之前没有消费该分区数据,因此消费偏移量为 0。

```
=========主题:kafkatopic2
分区:0
偏移量:1
key:kafkatopic2
消息内容:Send Kafka Message2
创建消息时间戳:1682348468380
线程:org.springframework.kafka.KafkaListenerEndpointContainer#0-0-C-1
=========主题:kafkatopic1
分区:0
偏移量:1
```

```
key:kafkatopic1
消息内容:Send Kafka Message1
创建消息时间戳:1682348468380
线程:org.springframework.kafka.KafkaListenerEndpointContainer#0-1-C-1
=========主题:kafkatopic2
分区:1
偏移量:0
key:kafkatopic2
消息内容:Send Kafka Message2
创建消息时间戳:1682348468380
线程:org.springframework.kafka.KafkaListenerEndpointContainer#0-0-C-1
=========主题:kafkatopic1
分区:1
偏移量:0
key:kafkatopic1
消息内容:Send Kafka Message1
创建消息时间戳:1682348468380
线程:org.springframework.kafka.KafkaListenerEndpointContainer#0-1-C-1
```

4. 多线程指定消费起始偏移量

生产者向 Kafka 发送的消息除了被消费者组直接消费外,还可以被缓存起来,缓存时间可通过 Kafka 的 config 目录下 server.properties 文件内的 log.retention.hours 配置项设置,默认为 7 天(等于 168 小时),存储超过 7 天的消息会被自动清除。在缓存时间内,如果消费者组宕机,生产者继续向主题发送消息,消费者组恢复上线后仍然可以从宕机时的消费偏移量接着消费,保证数据不丢失。同时 Kafka 也允许我们通过 initialOffset 配置项自定义设置消费者组恢复上线后的消费偏移量,指定消费者组恢复上线后先从 initialOffset 处消费所有消息,然后消费新的消息。

在 KafkaConsumer 中修改原有的@KafkaListener 注解内容如下:让消费者组启动后,先从设定的各主题分区的 initialOffset 处消费消息。设定先消费 kafkatopic1 主题 0 分区且消费偏移量为 1 的消息、kafkatopic1 主题 1 分区且消费偏移量为 0 的消息、kafkatopic2 主题 0 分区且消费偏移量为 1 的消息、kafkatopic2 主题 1 分区且消费偏移量为 0 的消息。

```
@KafkaListener(topicPartitions = {
    @TopicPartition(topic = "kafkatopic1", partitionOffsets = {
        @PartitionOffset(partition = "0",initialOffset = "1"),
        @PartitionOffset(partition = "1",initialOffset = "0")}),
    @TopicPartition(topic = "kafkatopic2", partitionOffsets = {
        @PartitionOffset(partition = "0",initialOffset = "1"),
        @PartitionOffset(partition = "1",initialOffset = "0")})
},groupId = "kafkaconsumergroup",concurrency = "2")
```

启动 Springdemo7Application,等待 5000 毫秒后,在控制台可以看到,消费者组 kafkaconsumergroup 启动后会先根据 initialOffset 消费 kafkatopic1 和 kafkatopic2 主题各分区的历史消息,消费情况和@KafkaListener 注解设定的一致。

```
=========主题:kafkatopic1
分区:0
消费偏移量:1
key:kafkatopic1
消息内容:Send Kafka Message1
创建消息时间戳:1682502632386
线程:org.springframework.kafka.KafkaListenerEndpointContainer#0-0-C-1
=========主题:kafkatopic2
分区:0
消费偏移量:1
key:kafkatopic2
消息内容:Send Kafka Message2
创建消息时间戳:1682502632386
线程:org.springframework.kafka.KafkaListenerEndpointContainer#0-1-C-1
=========主题:kafkatopic2
分区:1
消费偏移量:0
key:kafkatopic2
消息内容:Send Kafka Message2
创建消息时间戳:1682502753746
线程:org.springframework.kafka.KafkaListenerEndpointContainer#0-1-C-1
=========主题:kafkatopic1
分区:1
消费偏移量:0
key:kafkatopic1
消息内容:Send Kafka Message1
创建消息时间戳:1682502632386
线程:org.springframework.kafka.KafkaListenerEndpointContainer#0-0-C-1
```

等待 5000 毫秒,待 KafkaProducer 生产者生产消息到达后,kafkaconsumergroup 再消费 kafkatopic1 和 kafkatopic2 主题各分区新到达的消息。

```
=========主题:kafkatopic1
分区:0
消费偏移量:2
key:kafkatopic1
消息内容:Send Kafka Message1
创建消息时间戳:1682522043797
线程:org.springframework.kafka.KafkaListenerEndpointContainer#0-0-C-1
=========主题:kafkatopic1
分区:1
消费偏移量:1
key:kafkatopic1
消息内容:Send Kafka Message1
创建消息时间戳:1682522043797
线程:org.springframework.kafka.KafkaListenerEndpointContainer#0-0-C-1
=========主题:kafkatopic2
分区:0
消费偏移量:2
```

```
key:kafkatopic2
消息内容:Send Kafka Message2
创建消息时间戳:1682522043797
线程:org.springframework.kafka.KafkaListenerEndpointContainer#0-1-C-1
=========主题:kafkatopic2
分区:1
消费偏移量:1
key:kafkatopic2
消息内容:Send Kafka Message2
创建消息时间戳:1682522043797
线程:org.springframework.kafka.KafkaListenerEndpointContainer#0-1-C-1
```

任务 7.3 综合案例：Kafka 采集主机运行信息

在本项目的任务 2 中我们演示了利用 Kafka 的传递字符串类的消息。然而在实际应用场景中更多会传递一些复杂的消息，如传递集合或 Java 对象形式的消息。本节将以一个综合案例——Kafka 对主机运行信息的采集、传递和存储，演示如何利用 Kafka 传递 Java 对象类型的信息。

7.3.1 案例任务

在运维监控系统中，通常会定时采集服务器运行信息，监控服务器性能，以便运维。本任务模仿运维监控场景下的数据采集和存储功能，任务要求实时采集本机运行信息（包括主机名、系统版本号、CPU 使用情况、内存使用情况等），并利用 Kafka 生产消费模型进行消息传递，消费者端收到消息后存入 Redis 数据库，同时 Redis 数据库中存储的数据也能被正常读取。在消息传递过程中，主机运行信息必须采用自定义 Java 对象封装并传递。

7.3.2 案例分析

该任务将 Kafka 消息传递和 Redis 数据存取相结合，编程思路为：首先需利用 Java 自带的操作系统的管理接口类 OperatingSystemMXBean 定时采集本机运行信息，将采集的信息以 Java 对象形式封装发送给 Kafka；其次，Kafka 消费者端收到数据后存入 Redis 数据库；最后，读取 Redis 数据库数据，校验比对数据是否正确。该任务基于 SpringDemo7 项目实现，任务的实现思路如下。

（1）任务中利用 Kafka 传递 Java 对象，利用 Redis 存储 Java 对象。需要引入 Json 对 Java 对象进行序列化和反序列化操作，因此，需在 pom.xml 文件中添加 lombok、Json 和 Redis 依赖。

（2）修改 Kafka 相关配置以方便传递 Java 对象，修改 Redis 相关配以方便存取 Java 对象。

（3）编程实现利用 OperatingSystemMXBean 对象定时采集本机运行信息，并以指定的主题发送给 Kafka。

(4) 编写 Kafka 消费者监听该主题获取消息,并将获取的消息存入 Redis。

(5) 读取 Redis 消息,校验数据是否正确。

7.3.3 任务实施

由于案例中主机运行信息封装在 Java 对象中,Kafka 数据的生产消费和 Redis 数据的存储和读取都涉及传递 Java 对象。传递 Java 对象需要进行序列化和反序列化操作,一般使用 Json 工具对 Java 对象进行序列化和反序列化操作。因此需引入 Json 依赖,这里直接引入 spring-boot-starter-web 依赖,spring-boot-starter-web 会自动引入 Spring Boot 自带的 Json 工具包 com.fasterxml.jackson。为简化 Java 实体类的编程,还需引入 lombok 依赖。与 Redis 交互,还需引入 spring-boot-starter-data-redis 和 commons-pool2 依赖。综上所述,pom.xml 文件需添加的具体依赖如下。

```xml
<dependency>
    <groupId>org.springframework.boot</groupId>
    <artifactId>spring-boot-starter-web</artifactId>
</dependency>
<dependency>
    <groupId>org.projectlombok</groupId>
    <artifactId>lombok</artifactId>
</dependency>
<dependency>
    <groupId>org.springframework.boot</groupId>
    <artifactId>spring-boot-starter-data-redis</artifactId>
</dependency>
<dependency>
    <groupId>org.apache.commons</groupId>
    <artifactId>commons-pool2</artifactId>
</dependency>
```

在 application.yaml 文件中添加有关 Kafka 的配置。因为主机运行信息封装在 Java 对象中,涉及 Kafka 传递 Java 对象,默认 Kafka 生产者向消费者传递字符串形式消息,序列化和反序列化机制只适用于字符串。而本案例中 Kafka 生产者需要向消费者传递 Java 对象形式消息,就需要重新设置 Kafka 序列化和反序列化机制。这里设置生产者向消费者传递 Java 对象时序列化和反序列化使用 JsonSerializer 和 JsonDeserializer 类,即生产者传递消息将 Java 对象序列化成 Json 字符串,消费者消费消息时又利用 JsonDeserializer 类将 Json 字符串转为 Java 对象。

注意还需配置 properties.spring.json.trusted.packages 项,将 springdemo7.domain 目录下的所有 Java 类设为可信任的。因为 JsonDeserializer 在反序列化消息时要考虑到安全性,要反序列化成信任的 Java 类。如果不配置,项目启动时会报错。Redis 相关配置均使用默认值,无须额外配置。

```yaml
kafka:
  consumer:
    #消费者组关键字的反序列化类,默认用反序列化字符串
```

```yaml
      key-deserializer: org.apache.kafka.common.serialization.StringDeserializer
      #消费者组值的反序列化类,Json字符串经反序列化变为Java对象
      value-deserializer: org.springframework.kafka.support.serializer.JsonDeserializer
      properties:
        spring:
          json:
            trusted:
              packages: springdemo7.domain
    producer:
      #生产者关键字的序列化类,默认序列化成字符串
      key-serializer: org.apache.kafka.common.serialization.StringSerializer
      #生产者值的序列化类,Java对象序列化为Json字符串
      value-serializer: org.springframework.kafka.support.serializer.JsonSerializer
```

在 config 目录下新建 RedisConfig 配置类,配置向 Redis 传递 Java 对象时,使用 com.fasterxml.jackson 类对 Java 对象进行序列化和反序列化操作。

【RedisConfig.java】

```java
package springdemo7.config;
import com.fasterxml.jackson.annotation.JsonAutoDetect;
import com.fasterxml.jackson.annotation.PropertyAccessor;
import com.fasterxml.jackson.databind.ObjectMapper;
import com.fasterxml.jackson.databind.jsontype.impl.LaissezFaireSubTypeValidator;
import org.springframework.context.annotation.Bean;
import org.springframework.context.annotation.Configuration;
import org.springframework.data.redis.connection.lettuce.LettuceConnectionFactory;
import org.springframework.data.redis.core.RedisTemplate;
import org.springframework.data.redis.serializer.Jackson2JsonRedisSerializer;
import org.springframework.data.redis.serializer.StringRedisSerializer;
@Configuration
public class RedisConfig {
    @Bean
    public RedisTemplate<String, Object> redisTemplate(
        LettuceConnectionFactory lettuceConnectionFactory) {
            RedisTemplate<String, Object> redisTemplate = new RedisTemplate<>();
            redisTemplate.setConnectionFactory(lettuceConnectionFactory);
            ObjectMapper objectMapper = new ObjectMapper();
            //指定类中要序列化的范围
            objectMapper.setVisibility(PropertyAccessor.ALL, JsonAutoDetect.Visibility.ANY);
            //序列化时将对象全类名一起保存下来,以便反序列化
            objectMapper.activateDefaultTyping(LaissezFaireSubTypeValidator.instance, ObjectMapper.DefaultTyping.NON_FINAL);
            //使用Jackson2JsonRedisSerializer来序列化和反序列化redis的value值
        Jackson2JsonRedisSerializer jackson2JsonRedisSerializer =
                new Jackson2JsonRedisSerializer(objectMapper,Object.class);
        //设置key和value的序列化规则
        redisTemplate.setKeySerializer(new StringRedisSerializer());
        redisTemplate.setValueSerializer(jackson2JsonRedisSerializer);
```

```
        //设置hashKey和hashValue的序列化规则
        redisTemplate.setHashKeySerializer(new StringRedisSerializer());
        redisTemplate.setHashValueSerializer(jackson2JsonRedisSerializer);
        redisTemplate.afterPropertiesSet();
        return redisTemplate;
    }
}
```

在springdemo7目录下新建domain目录,在domain目录新建MachineInfo类并用于封装主机运行信息。在MachineInfo类上添加lombok的@Data、@AllArgsConstructor和@NoArgsConstructor注解。

【MachineInfo.java】

```
package springdemo7.domain;
import lombok.AllArgsConstructor;
import lombok.Data;
import lombok.NoArgsConstructor;
@Data
@AllArgsConstructor
@NoArgsConstructor
public class MachineInfo {
    private String name;                              //主机名
    private String version;                           //版本名
    private String arch;                              //处理器架构
    private Integer cpuCore;                          //CPU核数
    private String systemCpuLoad;                     //系统CPU使用率
    private String processCpuLoad;                    //进程CPU使用率
    private String processCpuTime;                    //进程使用CPU时间
    private String totalMemorySize;                   //总物理内存数量
    private String freeMemorySize;                    //空闲物理内存数量
    private String memoryUseRatio;                    //物理内存使用率
    private String committedVirtualMemorySize;        //已提交虚拟内存数量
    private String totalSwapSpaceSize;                //总交换空间数量
    private String freeSwapSpaceSize;                 //空闲交换空间数量
}
```

在producers目录下新建MachineKafkaProducer类,用于向Kafka发送实时采集的主机运行信息,主机运行信息采用Java对象封装。

【MachineKafkaProducer.java】

```
package springdemo7.producers;
import org.springframework.beans.factory.annotation.Autowired;
import org.springframework.kafka.core.KafkaTemplate;
import org.springframework.stereotype.Component;
import springdemo7.domain.MachineInfo;
@Component
public class MachineKafkaProducer {
    @Autowired(required = false)
```

```
    private KafkaTemplate kafkaTemplate;
    public void send(MachineInfo machineInfo) {
        //发送主题为 Machine 的消息
        kafkaTemplate.send("Machine",machineInfo);
    }
}
```

在 consumers 目录下新建 MachineKafkaConsumer 类,用于消费 Kafka 发送的实时主机运行信息,并存入 Redis。

【MachineKafkaConsumer.java】

```
package springdemo7.consumers;
import org.apache.kafka.clients.consumer.ConsumerRecord;
import org.springframework.beans.factory.annotation.Autowired;
import org.springframework.data.redis.core.RedisTemplate;
import org.springframework.kafka.annotation.KafkaListener;
import org.springframework.kafka.annotation.TopicPartition;
import org.springframework.stereotype.Component;
import java.time.LocalDateTime;
import java.time.format.DateTimeFormatter;
@Component
public class MachineKafkaConsumer {
    @Autowired
    private RedisTemplate redisTemplate;
    @KafkaListener(topicPartitions = {
            @TopicPartition(topic = "Machine",partitions = {"0"})
    },groupId = "kafkaconsumermachine")
    public void receiveMsg(ConsumerRecord consumerRecord) {
        String nowtime= LocalDateTime.now().format(
                DateTimeFormatter.ofPattern("yyyy-MM-dd HH:mm:ss"));
        //Kafka 消费 5 条主机运行信息并存入 Redis,键为时间,值为 MachineInfo 对象
        redisTemplate.opsForValue().set(nowtime,consumerRecord.value());
    }
}
```

在 springdemo7 目录下新建 MachineInfoTask 测试类,该类内部实现的 ApplicationListener 接口监听 ApplicationReadyEvent 事件。在项目启动后,该事件利用 Java 自带的操作系统的管理接口类 OperatingSystemMXBean 每隔 5 秒采集一次主机运行信息并发送 Kafka 生产者,共采集 3 次。然后休眠 2 秒后,读取并打印 Redis 中存储的主机运行信息。

【MachineInfoTask.java】

```
package springdemo7;
import com.fasterxml.jackson.databind.ObjectMapper;
import org.springframework.beans.factory.annotation.Autowired;
import org.springframework.boot.context.event.ApplicationReadyEvent;
import org.springframework.context.ApplicationListener;
import org.springframework.data.redis.core.RedisTemplate;
```

```java
import org.springframework.stereotype.Component;
import java.lang.management.ManagementFactory;
import java.time.LocalDate;
import java.time.format.DateTimeFormatter;
import java.util.Set;
import com.sun.management.OperatingSystemMXBean;
import springdemo7.domain.MachineInfo;
import springdemo7.producers.MachineKafkaProducer;
@Component
public class MachineInfoTask implements
        ApplicationListener<ApplicationReadyEvent> {
    @Autowired
    private MachineKafkaProducer machineKafkaProducer;
    @Autowired
    private RedisTemplate redisTemplate;
    final long unit = 1024 * 1024 * 1024;
    int  count=0;
    @Override
    public void onApplicationEvent(ApplicationReadyEvent event){
        try{
            //每隔 5 秒采集一次主机运行信息并发送 Kafka 生产者,共采集 3 次
            while(count<3) {
                OperatingSystemMXBean operatingSystemMXBean =
                    (OperatingSystemMXBean) ManagementFactory.
                        getOperatingSystemMXBean();
                MachineInfo machineInfo = new MachineInfo();
                machineInfo.setName(operatingSystemMXBean.getName());
                machineInfo.setVersion(operatingSystemMXBean.getVersion());
                machineInfo.setArch(operatingSystemMXBean.getArch());
                int CpuCore = operatingSystemMXBean.getAvailableProcessors();
                machineInfo.setCpuCore(CpuCore);
                Double SystemCpuLoad =
                        operatingSystemMXBean.getCpuLoad();
                machineInfo.setSystemCpuLoad(SystemCpuLoad * 100 + "%");
                Double ProcessCpuLoad =
                        operatingSystemMXBean.getProcessCpuLoad();
                machineInfo.setProcessCpuLoad(ProcessCpuLoad * 100 + "%");
                Long ProcessCpuTime = operatingSystemMXBean.getProcessCpuTime();
                machineInfo.setProcessCpuTime(ProcessCpuTime/1000000000.0+"秒");
                Long TotalMemorySize =
                        operatingSystemMXBean.getTotalMemorySize();
                machineInfo.setTotalMemorySize(TotalMemorySize / unit + "GB");
                Long FreeMemorySize = operatingSystemMXBean.getFreeMemorySize();
                machineInfo.setFreeMemorySize(FreeMemorySize / unit + "GB");
                Double MemoryUseRatio = (double)100 * (TotalMemorySize-
                        FreeMemorySize)/TotalMemorySize;
                machineInfo.setMemoryUseRatio(MemoryUseRatio + "%");
                Long CommittedVirtualMemorySize = operatingSystemMXBean.
                        getCommittedVirtualMemorySize();
                machineInfo.setCommittedVirtualMemorySize(
```

```
                CommittedVirtualMemorySize/unit+"GB");
            Long TotalSwapSpaceSize = operatingSystemMXBean.
                getTotalSwapSpaceSize();
            machineInfo.setTotalSwapSpaceSize(
                TotalSwapSpaceSize/unit+"GB");
            Long FreeSwapSpaceSize =
                operatingSystemMXBean.getFreeSwapSpaceSize();
            machineInfo.setFreeSwapSpaceSize(FreeSwapSpaceSize/unit+"GB");
            System.out.println(machineInfo);
            machineKafkaProducer.send(machineInfo);
            Thread.sleep(5000);
            count++;
        }
        Thread.sleep(2000);
        System.out.println("查询 Redis 中的数据=======");
        String nowdate= LocalDate.now().format(
            DateTimeFormatter.ofPattern("yyyy-MM-dd"));
        //模糊匹配获取 Redis 中键为今天的所有记录
        Set<String> keys = redisTemplate.keys(nowdate+"*");
        if(keys.size()!=0){
            ObjectMapper objectMapper = new ObjectMapper();
            keys.stream().forEach(elem->{
                //依次获取所有的 MachineInfo 对象并打印输出
                MachineInfo machineInfo = (MachineInfo) redisTemplate.
                    opsForValue().get(elem);
                if(null!=machineInfo){
                    System.out.println(machineInfo);
                }
            });
        }
    }catch(InterruptedException e){
        e.printStackTrace();
    }
  }
}
```

为了减少控制台干扰信息,这里先注释掉 KafkaProduce、KafkaConsumer 和 KafkaTest 上面的@Component 注解,使之都不运行。然后再启动 MachineInfoTask 类,控制台可以看到如下信息。MachineInfoTask 类采集了 3 条数据存入 Redis,随后成功读取了 Redis 存储的 3 条数据,数据完全一致。

```
MachineInfo(name=Windows 10,version=10.0,arch=amd64,cpuCore=8,
systemCpuLoad=10.040293072592688%,processCpuLoad=1.33409380888555285%,
processCpuTime=8.21875 秒,totalMemorySize=31GB,freeMemorySize=24GB,
memoryUseRatio=22.28798745787845%,committedVirtualMemorySize=0GB,
totalSwapSpaceSize=36GB,freeSwapSpaceSize=20GB)
MachineInfo(name=Windows 10,version=10.0,arch=amd64,cpuCore=8,
systemCpuLoad=9.315739689261859%,processCpuLoad=0.077590086186681945%,
processCpuTime=8.25 秒,totalMemorySize=31GB,freeMemorySize=24GB,
```

```
memoryUseRatio=22.308241055949935%,committedVirtualMemorySize=0GB,
totalSwapSpaceSize=36GB,freeSwapSpaceSize=20GB)
MachineInfo(name=Windows 10,version=10.0,arch=amd64,cpuCore=8,
systemCpuLoad=13.803791692572654%,processCpuLoad=0.2712526790216593%,
processCpuTime=8.359375秒,totalMemorySize=31GB,freeMemorySize=24GB,
memoryUseRatio=22.288407159437067%,committedVirtualMemorySize=0GB,
totalSwapSpaceSize=36GB,freeSwapSpaceSize=20GB)
查询Redis中的数据=======
MachineInfo(name=Windows 10,version=10.0,arch=amd64,cpuCore=8,
systemCpuLoad=10.040293072592688%,processCpuLoad=1.3340938088855285%,
processCpuTime=8.21875秒,totalMemorySize=31GB,freeMemorySize=24GB,
memoryUseRatio=22.28798745787845%,committedVirtualMemorySize=0GB,
totalSwapSpaceSize=36GB,freeSwapSpaceSize=20GB)
MachineInfo(name=Windows 10,version=10.0,arch=amd64,cpuCore=8,
systemCpuLoad=13.803791692572654%,processCpuLoad=0.2712526790216593%,
processCpuTime=8.359375秒,totalMemorySize=31GB,freeMemorySize=24GB,
memoryUseRatio=22.288407159437067%,committedVirtualMemorySize=0GB,
totalSwapSpaceSize=36GB,freeSwapSpaceSize=20GB)
MachineInfo(name=Windows 10,version=10.0,arch=amd64,cpuCore=8,
systemCpuLoad=9.315739689261859%,processCpuLoad=0.07759008618681945%,
processCpuTime=8.25秒,totalMemorySize=31GB,freeMemorySize=24GB,
memoryUseRatio=22.308241055949935%,committedVirtualMemorySize=0GB,
totalSwapSpaceSize=36GB,freeSwapSpaceSize=20GB)
```

Redis中存储的数据如图7-3所示。

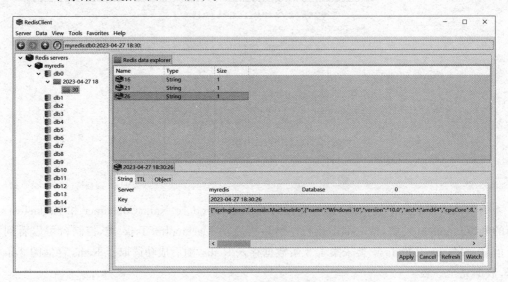

图7-3 Redis中存储的主机运行数据

小　　结

本项目详细介绍了基于Spring Boot项目的消息队列编程,包括Kafka基于主题、分区和偏移量的多种不同消费模式的编程实现。最后以一个综合案例——Kafka主机运行信息的采集、传递和存储,演示如何在 Spring Boot 项目中综合运用 Kafka 生产消费模型传递 Java 对象数据,并结合 Redis 进行数据存取,使读者进一步掌握 Spring Boot 消息队列的编程。

课后练习: Kafka 采集键盘输入字符数据

在键盘中输入一些字符数据,用 Kafka 编程,实时采集键盘输入的字符数据并发送给消费者端;消费者端能够实时接收 Kafka 传递的输入字符数据,并在控制台打印出来。

项目 8 Spring Boot Web 应用开发——后端

Web 应用是 Java 开发的一个重要方向。Spring Boot 默认使用 Spring MVC 框架开发 Web 应用的后端代码。Spring Boot 作为 Spring 框架的一个模板项目,基于约定优于配置的原则,能够最大化减少 Spring MVC 的使用成本,简化 Web 应用开发。

任务 8.1 初识 Spring MVC

在使用 Spring MVC 之前,需先了解 Spring MVC。本任务主要介绍 Spring MVC 的相关理论知识,包括什么是 Spring MVC、Spring MVC 的工作流程、Spring MVC 的引入和 Spring MVC 的单元测试。

8.1.1 Spring MVC 简介

Spring MVC 是 Spring 提供的一款以 MVC 设计模式为核心的轻量级 Web 框架,是 Spring 框架的一部分。与传统的 Web 框架相比,Spring MVC 学习成本更低,性能更加高效,能够更方便地结合 Spring 生态系统中的其他框架开发企业级 Web 应用系统。

MVC 是模型(model)、视图(view)、控制器(controller)的简写,是一种软件设计规范,通过将业务逻辑、数据、显示分离的方法来组织代码。在传统的 MVC 三层架构中,控制器一般使用 Servlet 实现,一个请求对应一个 Servlet。如果一个应用非常复杂,则需要同时注册编写多个 Servlet 类,这非常不方便。

Spring MVC 基于请求响应模式,针对传统 MVC 三层架构中的 Controller 层进行优化,以一个默认的 Servlet(DispatcherServlet)统一接收所有客户端请求并解析,再根据请求 URL 转发给不同的 Java 类处理,这些 Java 类都必须实现 Controller 接口,称为控制器类。处理完毕,结果统一发回 DispatcherServlet,由 DispatcherServlet 选择具体的页面进行渲染;渲染完毕,将渲染结果发送给客户端。Spring MVC 本质上就是一个 Servlet。与传统 MVC 三层架构相比,使用 Spring MVC 后,一个 Web 项目通常只需要一个 Servlet 就可以统一管理所有的请求和响应,减少项目的复杂度和开发成本。同时 Spring MVC 将 Model 层划分为业务逻辑层和数据持久层,减少了数据访问层和业务逻辑层的耦合度,使整个 Web 项目的架构更加合理。

8.1.2　Spring MVC 工作流程

下面简单介绍 Spring MVC 的工作流程。Spring MVC 的工作流程如图 8-1 所示。

Spring MVC 工作流程

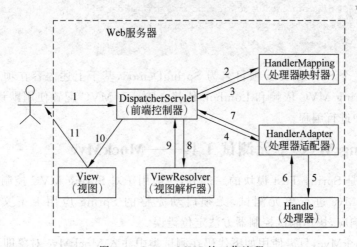

图 8-1　Spring MVC 的工作流程

基于图 8-1，Sping MVC 总体工作流程可分为如下步骤。

（1）用户从客户端发出请求，DispatcherServlet 统一拦截所有请求。

（2）DispatcherServlet 拦截请求后，调用处理器映射器 HandlerMapping。

（3）HandlerMapping 根据请求 URL 查找相应的 Handler(实现 Controller 接口的 Java 类)，并将结果返回给 DispatcherServlet。

（4）DispatcherServlet 根据结果选择特定的 HandlerAdapter。

（5）HandlerAdapter 选择相应的 Handler 处理请求。

（6）Handler 处理完毕，将具体的响应信息封装成对象，返回给 HandlerAdapter。最常用的是 ModelAndView 对象，内部包含模型数据和视图名。

（7）HandlerAdapter 将 ModelAndView 对象发送给 DispatcherServlet。

（8）DispatcherServlet 调用 ViewResolver 解析 ModelAndView 对象中的视图名。

（9）ViewResolver 将解析的视图名传给 DispatcherServlet。

（10）DispatcherServlet 根据 ModelAndView 中的模型数据渲染视图页面。

（11）渲染完毕，将视图页面发送给客户端。

使用 Spring MVC 时，开发人员不需要编程实现 DispatcherServlet、HandlerMapping、HandlerAdapter 和 ViewResolver 这些对象。只需配置默认的 DispatcherServlet，创建 Java 类并继承 Controller 接口，在 Java 类内部编程完成业务逻辑处理并在相应的页面视图展现即可。

8.1.3　Spring Boot 引入 Spring MVC

　　Spring Boot 和 Spring MVC 都是 Spring 框架的一部分，二者可以方便地整合。在整合过程中，Spring Boot 采用约定优于配置的原则，通过 WebMvcAutoConfiguration 类对 Spring MVC 进行自动配置，开发人员不需额外配置，就可以直接使用 Spring MVC 开发 Web 应用了。在 Spring Boot 项目中使用 Spring MVC，只需在项目的 pom.xml 文件中引入如下 Web 依赖。

```xml
<dependency>
    <groupId>org.springframework.boot</groupId>
    <artifactId>spring-boot-starter-web</artifactId>
</dependency>
```

　　这里新建一个 Spring Boot 项目名为 SpringDemo8，基于上述内容在项目的 pom.xml 文件中引入 Spring MVC 依赖和 Lombok 依赖。Spring MVC 配置使用默认配置，后续案例将在该项目中进行编程。

8.1.4　Spring MVC 单元测试工具——MockMvc

　　MockMvc 是 Spring Test 模块的一部分，一般用于对 Spring MVC 控制器进行单元测试。使用 MockMvc 进行单元测试，无须启动完整的 Spring 应用上下文，就可以模拟 HTTP 请求和响应，快速测试控制器方法定位错误。

　　要使用 MockMvc，只需使用如下代码在测试类中注入 MockMvc 对象即可。

```
@Autowired
private MockMvc mockMvc;
```

　　注入 MockMvc 对象后，就可以使用 mockMvc.perform 方法模拟 HTTP 请求了。

1. MockMvc 发送请求

　　如下代码就模拟发送了一个带参数的 get 请求。

```
mockMvc.perform(get("/path").param("param","value"));
```

　　其中，perform() 方法内部通过 get("/path") 设置 get 请求路径，通过 .param("param", "value") 封装了请求参数。如有多个参数，可以编写多个 .param("param","value") 封装数据。

　　如下代码又模拟了一个带参数的 post 请求。因为 post 请求的数据放在请求体中，因此请求参数封装在 content 中。如有多个参数，这些参数需以 param1＝value1＆param2＝value2 形拼接在一起放入 content 中。

```
mockMvc.perform(post("/path").content("request body data"));
```

2. MockMvc 获取响应

　　mockMvc.perform 方法执行后返回 ResultActions 对象，内部封装响应信息。如需获取响应信息，可以使用 ResultActions 对象提供的相关方法。如可采用如下写法获取

HTTP 响应状态码是否为 200。

```
ResultActions action = mvc.perform(get("/path"));
action.andExpect(status().isOk())      //获取 HTTP 状态码是否为 200
```

如果要获取 HTTP 响应数据内容，可使用如下写法获取。

```
ResultActions action = mvc.perform(get("/path"));
action.andExpect(status().isOk());
//获取请求处理方法的返回值
String res = action.andReturn().getResponse().getContentAsString();
```

上述内容介绍一些 MockMvc 的初步使用，MockMvc 的更多应用将在后续内容中展示。

任务 8.2　Spring MVC 访问静态资源

在 Web 开发中少不了要访问一些静态资源，例如首页文件、静态页面文件、CSS 文件、JavaScript 文件等。Spring Boot 内部定义的 CLASSPATH_RESOURCE_LOCATIONS 属性值用于设定 SpringMVC 静态资源访问目录。CLASSPATH_RESOURCE_LOCATIONS 是一个字符串数组，数组内部定义了如下 4 个静态地址。

```
private static final String[] CLASSPATH_RESOURCE_LOCATIONS = new String[]{
        "classpath:/META-INF/resources/",
        "classpath:/resources/",
        "classpath:/static/",
        "classpath:/public/"};
```

在编程时，这 4 个地址都可以用来放静态文件。其中，classpath 指向当前项目的根目录 resources。resources 根目录下自带 static 目录，可直接放置静态资源；如需使用其他 3 个目录放置静态资源，需手动创建。4 个地址按顺序优先级从高到低。其中，/META-INF/resources/目录优先级最高，/public/目录优先级最低。假设 4 个地址下都存在 a.css 文件，在浏览器中输入地址 http://localhost:8080/a.css，SpringMVC 会将/a.css 映射成/META-INF/resources/a.css 访问资源。

如果需输入项目根目录地址 http://localhost:8080/，浏览器直接跳转到首页 index.html，只需将 index.html 页面放在上述 4 个地址之一中即可，一般为方便，可以直接放在 static 目录下。

如需修改默认静态资源映射规则，可以在项目的 application.yaml 文件中修改如下的 static-path-pattern 配置，如将默认的拦截规则/**改为/aa/**，只拦截请求路径为当前项目/aa 的请求。

```
spring:
  mvc:
    static-path-pattern: /aa/**
```

此时在浏览器中输入地址 http://localhost:8080/a.css,但是无法访问 a.css,访问地址需改为 http://localhost:8080/aa/a.css。

如需从自定义目录加载静态资源,可以在项目的 application.yaml 文件中修改如下 static-locations 配置,如将静态资源目录改为 resources 目录下的 diy 文件夹。

```
spring:
  web:
    resources:
      static-locations: classpath:/diy/
```

在浏览器中输入地址 http://localhost:8080/a.css 并访问 a.css 文件,a.css 文件必须放置在 resources 的 diy 目录下才能正常访问。

任务 8.3 Spring MVC 访问动态资源 ——映射请求

Web 开发中除了访问静态资源外,更多的是访问动态资源。Spring MVC 通过默认控制器 DispatchServlet 统一处理对动态资源的请求访问,DispatchServlet 将请求根据地址分发给不同的控制器处理,控制器需由开发人员根据业务具体编码实现。控制器的编码实现主要涉及三方面内容,分别是如何映射请求、如何获取请求数据和如何输出响应。在映射请求方面,Spring MVC 可基于注解编程实现,与此相关的注解有 @Controller 注解、@RequestMapping 注解和组合注解等。本任务将介绍如何使用注解编程实现映射请求。

8.3.1 @Controller 注解

前面提到一个 Java 类要实现 Controller 接口,才会成为一个控制器类,才能处理 DispatchServlet 分发的请求。除此之外,Spring MVC 也提供基于注解方式来实现控制器。SpringMVC 提供的 @Controller 注解用于标识一个 Java 类为控制器类,该 Java 类内部可以定义多个方法分别处理不同路径的请求。当 Java 类上添加了 @Controller 注解,Spring 容器初始化时会自动扫描设定路径下所有添加 @Controller 注解的 Java 类,将这些 Java 类实例化成控制器对象。如果是 Spring Boot 项目,默认设定路径为根目录。

```
import org.springframework.stereotype.Controller;
@Controller
public class MyController {
}
```

8.3.2 @RequestMapping 注解

当一个 Java 类被 @Controller 标识成控制器后,内部就可以定义多个方法分别处理不同路径的请求,这就需要用到 @RequestMapping 注解。@RequestMapping 注解内部默认提供 value 属性用于设置请求路径和控制器内部方法的映射关系。在使用时,

@RequestMapping 注解可以标注在控制器内部方法上,也可以标注在控制器类上。下面具体介绍,这里假设在项目的 springdemo8 目录下新建 controller 目录。

1. @RequestMapping 注解标注在方法上

@RequestMapping 注解标注的方法处理@RequestMapping 注解对应的 URL 请求。在方法上使用@RequestMapping 注解的示例如下。

```
package springdemo8.controller;
import org.springframework.stereotype.Controller;
import org.springframework.web.bind.annotation.RequestMapping;
@Controller
public class MyController {
    @RequestMapping(value="/myController")
    public void method(){
        System.out.println("MyController.method process");
    }
}
```

其中,MyController 类的 method 方法用@RequestMapping(value="/myController")注解修饰。在浏览器中输入地址 http://localhost:8080/myController,method 方法将处理该请求,控制台打印 MyController.method process。

2. @RequestMapping 注解标注在类上

@RequestMapping 注解标注在类上时,该类内部所有方法处理的请求地址前都必须添加类上标注的@RequestMapping 注解地址,实现访问地址的模块化管理。@RequestMapping 注解标注在类上的示例如下。

```
package springdemo8.controller;
import org.springframework.stereotype.Controller;
import org.springframework.web.bind.annotation.RequestMapping;
@Controller
@RequestMapping(value="/myController")
public class MyController {
    @RequestMapping(value="/myController")
    public void method(){
        System.out.println("MyController.method process");
    }
}
```

其中,MyController 类用@RequestMapping(value="/myController")注解修饰且内部 method 方法也用@RequestMapping(value="/myController")注解修饰。在浏览器中输入地址 http://localhost:8080/myController/myController,method 方法将处理该请求,控制台打印 MyController.method process。

@RequestMapping 注解除了拥有 value 属性外,还有一些其他属性。@RequestMapping 注解中的所有属性都是可选的,如设置时不指定属性名,则默认匹配 value 属性。因此,@RequestMapping(value="/myController")又可写成@RequestMapping("/myController")。@RequestMapping 注解属性如表 8-1 所示。

表 8-1 @RequestMapping 注解属性

属性名	类型	描述
name	String	可选属性,为请求的地址设置一个别名
value	String[]	可选属性,设置处理请求的地址,支持数组形式设置多个地址
method	enum RequestMethod	可选属性,用于设置处理请求的类型,值为 RequestMethod 枚举类型 GET、HEAD、POST、PUT、PATCH、DELETE、OPTIONS、TRACE 中的一个或多个,常用的有 GET 和 POST 类型。如 method = RequestMethod.GET 表示该方法只拦截 GET 请求。国内一般较多使用 GET 和 POST 请求
params	String[]	可选属性,用于设置处理请求地址包含特定参数名的请求。一般和 value 结合使用,用于处理请求时获取请求参数值
headers	String[]	可选属性,用于设置处理特定请求头的请求
consumes	String[]	可选属性,用于设置处理特定内容类型(Content-type)的请求,如 consumes = "application/json"表示拦截内容为 JSON 格式的请求
producers	String[]	可选属性,用于设置处理返回值为特定类型和特定编码的请求,producers 设置的内容必须在 request 请求头 Accept 中包含

下面基于表 8-1 中的属性列举一些@RequestMapping 注解的常用示例。

(1) 处理地址为 http://localhost:8080/myController 的请求,别名为 alias_myController。即

```
@RequestMapping(value="/myController",name="/alias_myController")
```

(2) 处理地址为 http://localhost:8080/myController 和 http://localhost:8080/myController1。即

```
@RequestMapping(value={"/myController","/myController1"})
```

(3) 只处理地址为 http://localhost:8080/myController 的 GET 和 POST 请求。即

```
@RequestMapping(value="/myController",method={RequestMethod.GET,
RequestMethod.POST}
```

(4) 处理地址为 http://localhost:8080/myController? id=1&&name=2 的请求,请求参数必须包含 id 和 name。即

```
@RequestMapping(value="/myController",params = {"id","name"})
```

(5) 处理地址为 http://localhost:8080/myController 的请求,请求参数不包含 id。即

```
@RequestMapping(value="/myController",params = "!id")
```

(6) 处理地址为 http://localhost:8080/myController 的请求,请求头必须包含 headers。即

```
@RequestMapping(value = "/test",headers = "headers")
```

(7) 处理地址为 http://localhost:8080/myController 的请求,请求内容为 JSON 格式。即

```
@RequestMapping(value = "/myController", consumes="application/json")
```

(8) 处理地址为 http://localhost:8080/myController 的请求,请求返回内容类型为 JSON 格式且编码为 UTF8。即

```
@RequestMapping(value="myController",produces = {"application/json;charset=
UTF-8"})
```

8.3.3 组合注解

为进一步简化@RequestMapping 注解的使用,Spring 框架提供了组合注解@ * Mapping,将一些常用的请求方式和@RequestMapping 注解进行整合。例如@RequestMapping(method={RequestMethod.GET})可以整合成@GetMapping。常用的组合注解如下所示。

```
@GetMapping 用于处理 get 方式请求
@PostMapping 用于处理 post 方式请求
@PutMapping 用于处理 put 方式请求
@DeleteMapping 用于处理 delete 方式请求
@PatchMapping 用于处理 patch 方式请求
```

如在实际使用中,注解@RequestMapping(value="/aa",method={RequestMethod.GET})就可以利用组合注解@GetMapping(value="/aa")代替,减少了代码量。

任务 8.4　Spring MVC 访问动态资源——获取请求数据

在利用@RequestMapping 映射请求后,接下来就需要对请求进行处理。如果请求带有参数,Spring MVC 会将请求参数中的数据进行相应转换并根据特定的规则绑定到控制器类的方法参数中,开发人员就可以通过控制器方法参数获取请求数据了。Spring MVC 支持 Servlet 原生对象——HttpServletRequest 对象的 getParameter 方法获取请求数据。除此之外,Spring MVC 还提供了多种注解,它们以不同方式获取请求数据,这些注解统一放在 org.springframework.web.bind.annotation 包下,比较常用的注解有@RequestParam、@RequsetBody 和@PathVariable 等。下面将介绍如何使用这些注解获取请求数据。

8.4.1　@RequestParam 注解

@RequestParam 注解用于方法参数上,一般用于获取少量的请求数据。@RequestParam

允许开发人员自定义控制器方法参数名和请求参数名之间的数据绑定规则。@RequestParam 注解内部有 4 个属性,如表 8-2 所示。

表 8-2　@RequestParam 注解属性

属性名	类型	描　　述
name	String	可选属性,用于设定控制器方法参数绑定哪个请求参数获取数据
value	String	可选属性,作用和 name 一样,用于设定控制器方法参数绑定哪个请求参数获取数据。使用时 value 和 name 只需设置一个。单独使用该属性可以省略 value,如 @RequestMapping(value="/aaa")可以写成@RequestMapping("/aaa")
required	boolean	可选属性,用于指定请求时是否必须传入该参数,默认为 true,表示必须传入
defaultValue	String	可选属性,请求时如果没有传入该参数,使用自定义默认值绑定到控制器方法参数中,仅当 required=false 时该属性生效

@RequestParam 注解使用示例如下。

```
@RequestMapping(value="/myController")
public String method1(@RequestParam(value="user_id", required = false,
defaultValue = "0") Integer id){
    System.out.println("id="+id);
    return "success";
}
```

如果请求地址为 http://localhost:8080/myControlle? user_id=1,method 方法将通过参数 id 获取请求参数 user_id 的值,控制台输出 id=1;如果请求地址为 http://localhost:8080/myControlle,请求不带参数,则使用默认值 0 传给 method 方法的参数 id,控制台输出 id=0。在实际编码中,虽然@RequestParam 可以设定控制器方法参数名称和请求参数名称的对应关系,但是一般情况下还是建议让二者名称保持一致,减少代码量。

@RequestParam 用于获取 GET 或 POST 请求中格式为 xx=xx&xx=xx 的字符串数据。如请求方式为 GET,数据放置在请求头中无须设定 content-type;如请求方式为 POST,数据放置在请求体中必须设定 Content-Type 为 application/x-www-form-urlencoded。

@RequestParam 支持以 Integer、String 等简单数据类型获取数据,如请求有多个参数,可通过多个@RequestParam 依次获取,也可通过 Map 集合一次性获取。@RequestParam 还支持以数组形式获取数据,但不支持以自定义 Java 对象形式获取数据。如果控制器方法参数名和请求参数名完全一致,数据绑定可以省略@RequestParam,直接通过名称匹配绑定。@RequestParam获取数据的常用示例如下。

(1) @RequestParam 以 String 类型依次获取 GET 或 POST 请求数据。

```
method(@RequestParam String username,@RequestParam String password)
```

(2) @RequestParam 以 Map 集合一次性获取 GET 或 POST 请求数据。

```
method(@RequestParam Map map)
```

(3) @RequestParam 以整型数组方式获取 GET 或 POST 请求数据。

```
method(@RequestParam Integer[] id)
```

（4）省略@RequestParam，直接以参数名匹配，以字符串接收 GET 或 POST 请求数据。

```
method(String username,String password)
```

（5）省略@RequestParam，直接以参数名匹配，以自定义 User 对象接收 GET 或 POST 请求数据。

```
method(User user)
```

下面编程演示@RequestParam 注解的具体用法。

在 springdemo8 目录下新建 domain 目录，在 domain 目录下新建 User 类。

【User.java】

```java
package springdemo8.domain;
import lombok.AllArgsConstructor;
import lombok.Data;
import lombok.NoArgsConstructor;
@Data
@AllArgsConstructor
@NoArgsConstructor
public class User {
    private String username;
    private String password;
}
```

在 springdemo8 目录下新建 controller 目录，在 controller 目录下新建 UserController 类，内部定义的多个方法演示@RequestParam 的常用写法。

【UserController.java】

```java
package springdemo8.controller;
import org.springframework.stereotype.Controller;
import org.springframework.web.bind.annotation.RequestMapping;
import org.springframework.web.bind.annotation.RequestParam;
import springdemo8.domain.User;
import java.util.Arrays;
import java.util.Map;
@Controller
@RequestMapping("/user")
public class UserController {
    //以 String 类型获取 get 请求数据
    @RequestMapping(value="/getBasic")
    public void getBasic(@RequestParam String username,
                         @RequestParam String password){
        System.out.println("getBasic:username="+
```

```java
            username+"password="+password);
    }
    //以 Map 类型获取 get 请求数据
    @RequestMapping("/getMap")
    public void getMap(@RequestParam Map map){
        System.out.println("getMap:"+map);
    }
    //以整型数组类型获取 get 请求数据
    @RequestMapping("/getArray")
    public void getArray(@RequestParam Integer[] id){
        Arrays.stream(id).forEach(elem->{
            System.out.println("getArray:id="+elem);
        });
    }
    //以 String 类型获取 post 请求数据
    @RequestMapping(value="/postBasic")
    public void postBasic(@RequestParam String username,
                          @RequestParam String password){
        System.out.println("postBasic:username="+
                username+"password="+password);
    }
    //以 Map 类型获取 post 请求数据
    @RequestMapping(value="/postMap")
    public void postMap(@RequestParam Map map){
        System.out.println("postMap:"+map);
    }
    //以整型数组类型获取 post 请求数据
    @RequestMapping("/postArray")
    public void postArray(@RequestParam Integer[] id){
        Arrays.stream(id).forEach(elem->{
            System.out.println("postArray:id="+elem);
        });
    }
    //省略@RequestParam,以 String 类型获取 get 请求数据
    @RequestMapping(value="/getBasicNoRp")
    public void getBasicNoRp(String username,String password){
        System.out.println("getBasicNoRp:username="+
                username+"password="+password);
    }
    //省略@RequestParam,以 User 对象获取 get 请求数据
    @RequestMapping("/getUserNoRp")
    public void getUserNoRp(User user){
        System.out.println("getUserNoRp:"+user);
    }
}
```

生成 UserController 测试类文件 UserControllerTest。在 UserControllerTest 中利用 MockMvc 工具对 UserController 类中定义的方法进行测试。

【UserControllerTest.java】

```java
package springdemo8.controller;
import org.junit.jupiter.api.Test;
import org.springframework.beans.factory.annotation.Autowired;
import org.springframework.boot.test.autoconfigure.web.servlet.AutoConfigureMockMvc;
import org.springframework.boot.test.context.SpringBootTest;
import org.springframework.http.MediaType;
import org.springframework.test.web.servlet.MockMvc;
import org.springframework.test.web.servlet.ResultActions;
import org.springframework.test.web.servlet.request.MockHttpServletRequestBuilder;
import org.springframework.test.web.servlet.request.MockMvcRequestBuilders;
import org.springframework.test.web.servlet.result.MockMvcResultHandlers;
import org.springframework.test.web.servlet.result.MockMvcResultMatchers;
@SpringBootTest(webEnvironment = SpringBootTest.WebEnvironment.RANDOM_PORT)
@AutoConfigureMockMvc
class UserControllerTest {
    @Test
    void sendDataMain(@Autowired MockMvc mvc) throws Exception{
        System.out.println("以 String 类型获取 get 请求数据:");
        sendData(mvc,"get","/user/getBasic");
        System.out.println("以 Map 类型获取 get 请求数据:");
        sendData(mvc,"get","/user/getMap");
        System.out.println("以整型数组类型获取 get 请求数据:");
        sendData(mvc,"get","/user/getArray");
        System.out.println("以 String 类型获取 post 请求数据:");
        sendData(mvc,"post","/user/postBasic");
        System.out.println("以 Map 类型获取 post 请求数据:");
        sendData(mvc,"post","/user/postMap");
        System.out.println("以整型数组类型获取 post 请求数据");
        sendData(mvc,"post","/user/postArray");
    }
    void sendData(MockMvc mvc,String type,String url) throws Exception{
        MockHttpServletRequestBuilder builder;    //创建 HttpServletRequest 对象
        if(type.equals("get")){
            builder = MockMvcRequestBuilders.get(url);
            if(url.contains("Array")){
                builder .param("id","1")
                    .param("id","2");
            }else{
                builder .param("username","user1")
                    .param("password","password1");
            }
        }else{
            builder = MockMvcRequestBuilders.post(url)
                .contentType(MediaType.APPLICATION_FORM_URLENCODED_VALUE);
            if(url.contains("Array")){
                builder.content("id=1&id=2");
            }else{
                builder.content("username=user1&password=password1");
```

```
            }
        }
        //执行请求
        ResultActions action = mvc.perform(builder);
        //获取请求处理状态是否为 200
        action.andExpect(MockMvcResultMatchers.status().isOk())
                .andDo(MockMvcResultHandlers.print());
    }
}
```

运行测试类 UserControllerTest,控制台打印如下信息,所有方法都正常获取到请求数据了。

```
以 String 类型获取 get 请求数据:
getBasic:username=user1password=password1

以 Map 类型获取 get 请求数据:
getMap:{username=user1, password=password1}

以整型数组类型获取 get 请求数据:
getArray:id=1
getArray:id=2

以 String 类型获取 post 请求数据:
postBasic:username=user1password=password1

以 Map 类型获取 post 请求数据:
postMap:{username=user1, password=password1}

以整型数组类型获取 post 请求数据:
postArray:id=1
postArray:id=2
```

8.4.2 @RequsetBody 注解

@RequsetBody 注解用于方法参数上,一般用于获取请求体中的大批量数据。因为从请求体中获取数据,@RequsetBody 注解只能用于获取 POST 请求的字符串数据,数据可以为普通字符串或 JSON 格式字符串。@RequsetBody 注解内部只有一个 required 属性,用于指定@RequsetBody 注解修饰的变量值是否为空,默认值为 true,要求变量值不能为空。

与@RequestParam 注解相比,@RequsetBody 注解用法相对统一。@RequsetBody 注解最常用的方式是用来接收 JSON 格式的字符串数据,接收以后可以根据需要将数据转换为基本数据、数组、Map 对象、自定义 Java 对象和自定义 Java 对象集合等赋值给方法参数。一个请求处理方法中只能有一个@RequestBody 注解,但是可以有多个@RequestParam 注解。因此,@RequsetBody 注解数据绑定时方法参数名可以为任何字符,与请求参数没有直接关联。@RequsetBody 获取数据的常用示例如下。

(1) @RequsetBody 注解以 String 类型获取 POST 请求数据。即

```
method(@RequestBody String str)
```

(2) @RequsetBody 注解以 Map 集合获取 POST 请求数据。即

```
method(@RequestBody Map map)
```

(3) @RequsetBody 注解以整型数组方式获取 POST 请求数据。即

```
method(@RequestBody Integer[] id)
```

(4) @RequsetBody 注解以自定义 Java 对象方式获取 POST 请求数据。即

```
method(@RequestBody User user)
```

(5) @RequsetBody 注解以自定义 Java 对象集合方式获取 POST 请求数据。即

```
method(@RequestBody List<User> userList)
```

下面编程演示@RequsetBody 注解的具体用法。

在 controller 目录下新建 UserController1 类,内部定义多个方法,演示@RequestBody 的常用写法。

【UserController1.java】

```java
package springdemo8.controller;
import org.springframework.stereotype.Controller;
import org.springframework.web.bind.annotation.RequestBody;
import org.springframework.web.bind.annotation.RequestMapping;
import springdemo8.domain.User;
import java.util.Arrays;
import java.util.List;
import java.util.Map;
@Controller
@RequestMapping(value="/requestBody")
public class UserController1 {
    //以字符串获取 JSON 格式的 POST 请求数据
    @RequestMapping(value="/postJsonString")
    public void postJsonString(@RequestBody String str){
        System.out.println("jsonStr="+str);
    }
    //以 Map 获取 POST 请求数据
    @RequestMapping(value="/postMap")
    public void postMap(@RequestBody Map map){
        System.out.println("map="+map);
    }
    //以整型数组获取 POST 请求数据
    @RequestMapping(value="/postArray")
```

```java
    public void postArray(@RequestBody Integer[] id){
        Arrays.stream(id).forEach(elem->{
            System.out.println("postArray="+elem);
        });
    }
    //以 User 对象获取 POST 请求数据
    @RequestMapping(value="/postUserObject")
    public void postUserObject(@RequestBody User user){
        System.out.println("user="+user);
    }
    //以 User 对象集合获取 POST 请求数据
    @RequestMapping(value="/postUserList")
    public void postUserList(@RequestBody List<User> userList){
        System.out.println("userList="+userList);
    }
}
```

生成 UserController1 测试类文件 UserController1Test。在 UserController1Test 中利用 MockMvc 工具对 UserController1 类中定义的方法进行测试。

【UserController1Test.java】

```java
package springdemo8.controller;
import com.fasterxml.jackson.databind.ObjectMapper;
import org.junit.jupiter.api.Test;
import org.springframework.beans.factory.annotation.Autowired;
import org.springframework.boot.test.autoconfigure.web.servlet.AutoConfigureMockMvc;
import org.springframework.boot.test.context.SpringBootTest;
import org.springframework.http.MediaType;
import org.springframework.test.web.servlet.MockMvc;
import org.springframework.test.web.servlet.ResultActions;
import org.springframework.test.web.servlet.request.MockHttpServletRequestBuilder;
import org.springframework.test.web.servlet.request.MockMvcRequestBuilders;
import org.springframework.test.web.servlet.result.MockMvcResultMatchers;
import springdemo8.domain.User;
import java.util.ArrayList;
import java.util.HashMap;
import java.util.List;
import java.util.Map;
@SpringBootTest(webEnvironment = SpringBootTest.WebEnvironment.RANDOM_PORT)
@AutoConfigureMockMvc
class UserController1Test {
    @Test
    void sendDataMain(@Autowired MockMvc mvc) throws Exception{
        System.out.println("以字符串获取 JSON 格式的 POST 请求数据:");
        sendData(mvc,"/requestBody/postJsonString");
        System.out.println("以 Map 获取 POST 请求数据:");
        sendData(mvc,"/requestBody/postMap");
        System.out.println("POST 请求以整型数组获取数据:");
        sendData(mvc,"/requestBody/postArray");
```

```java
            System.out.println("以 User 对象获取 POST 请求数据:");
            sendData(mvc,"/requestBody/postUserObject");
            System.out.println("以 User 对象集合获取 POST 请求数据:");
            sendData(mvc,"/requestBody/postUserList");
    }
    void sendData(MockMvc mvc,String url)throws Exception{
        MockHttpServletRequestBuilder builder;   //创建 HttpServletRequest 对象
        ObjectMapper objectMapper=new ObjectMapper();
        if(url.contains("postJsonString")){
            Map map=new HashMap();
            map.put("name","name1");
            map.put("id",1);
            builder = MockMvcRequestBuilders.post(url)
                    .contentType(MediaType.APPLICATION_JSON_VALUE)
                    .content(objectMapper.writeValueAsString(map));
        }
        else if(url.contains("postMap")){
            Map map=new HashMap();
            map.put("name","name1");
            map.put("id",1);
            builder = MockMvcRequestBuilders.post(url)
                    .contentType(MediaType.APPLICATION_JSON_VALUE)
                    .content(objectMapper.writeValueAsString(map));
        }else if(url.contains("postArray")){
            List<Integer> intArray=new ArrayList<Integer>();
            intArray.add(1);
            intArray.add(2);
            builder = MockMvcRequestBuilders.post(url)
                    .contentType(MediaType.APPLICATION_JSON_VALUE)
                    .content(objectMapper.writeValueAsString(intArray));
        }else if(url.contains("postUserObject")){
            User user=new User("user1","password1");
            builder = MockMvcRequestBuilders.post(url)
                    .contentType(MediaType.APPLICATION_JSON_VALUE)
                    .content(objectMapper.writeValueAsString(user));
        }else{
            User user1=new User("user1","password1");
            User user2=new User("user2","password2");
            List<User> userList=new ArrayList<User>();
            userList.add(user1);
            userList.add(user2);
            builder = MockMvcRequestBuilders.post(url)
                    .contentType(MediaType.APPLICATION_JSON_VALUE)
                    .content(objectMapper.writeValueAsString(userList));
        }
        //执行请求
        ResultActions action = mvc.perform(builder);
        //获取请求处理状态是否为 200
        action.andExpect(MockMvcResultMatchers.status().isOk())
                .andDo(MockMvcResultHandlers.print());
    }
}
```

运行测试类 UserController1Test，控制台打印如下信息，所有方法都正常获取到请求数据了。

```
以字符串获取 JSON 格式的 POST 请求数据：
jsonStr={"name":"name1","id":1}
以 Map 获取 JSON 格式的 POST 请求数据：
map={name=name1, id=1}
以整型数组获取 JSON 格式的 POST 请求数据：
postArray=1
postArray=2
以 User 对象获取 JSON 格式的 POST 请求数据：
user=User(username=user1, password=password1)
以 User 对象集合获取 JSON 格式的 POST 请求数据：
userList=[User(username=user1, password=password1), User(username=user2,
assword=password2)]
```

8.4.3　@PathVariable 注解

@PathVariable 注解用于匹配拼接在 URL 地址中以 {xxx} 形式存在的一个或多个占位符参数，以此获取请求数据。@PathVariable 注解也是 SpringMVC 实现 Restful 风格请求的基础。@PathVariable 注解允许在 @RequestMapping 注解中编写带占位符的 URL 地址映射请求，@PathVariable 注解可以将 URL 中占位符参数 {xxx} 绑定到处理器类的方法形参中获取请求参数。如请求地址为 http://localhost:8080/1/aaa。@RequestMapping 注解中可以采用占位符参数方式编写 URL 映射地址。

```
@RequestMapping(value="user/{id}/{name}")
```

在请求处理方法中可以使用如下形式接收数据。其中 id 值为 1，name 值为 aaa。

```
(@PathVariable("id") Integer id,@PathVariable("name") String name)
```

由于请求数据是拼接在 URL 地址中通过占位符匹配获取的，@PathVariable 注解一般用于 GET 方式请求。与 @RequestParam 注解类似，@PathVariable 注解支持以 Integer、String 等简单数据类型获取数据，无法接收自定义 Java 对象类型数据，但是可以通过多个 @PathVariable 注解接收多个参数值。下面编程演示 @PathVariable 注解的具体用法。

在 controller 目录下新建 UserController2 类，内部定义一个方法，演示 @PathVariable 注解的常用写法。

【UserController2.java】

```
package springdemo8.controller;
import org.springframework.stereotype.Controller;
import org.springframework.web.bind.annotation.PathVariable;
import org.springframework.web.bind.annotation.RequestMapping;
@Controller
@RequestMapping("/pathVariable")
```

```
public class UserController2 {
    //@PathVariable注解接收多个简单数据类型数据
    @RequestMapping(value="/user/{username}/{password}")
    public void getString(@PathVariable String username,@PathVariable String password){
        System.out.println("username="+username+",password="+password);
    }
}
```

生成 UserController2 测试类文件 UserController2Test。在 UserController2Test 中利用 MockMvc 工具对 UserController2 类中定义的方法进行测试。

【UserController2Test.java】

```
package springdemo8.controller;
import org.junit.jupiter.api.Test;
import org.springframework.beans.factory.annotation.Autowired;
import org.springframework.boot.test.autoconfigure.web.servlet.AutoConfigureMockMvc;
import org.springframework.boot.test.context.SpringBootTest;
import org.springframework.test.web.servlet.MockMvc;
import org.springframework.test.web.servlet.ResultActions;
import org.springframework.test.web.servlet.request.MockHttpServletRequestBuilder;
import org.springframework.test.web.servlet.request.MockMvcRequestBuilders;
import org.springframework.test.web.servlet.result.MockMvcResultMatchers;
@SpringBootTest(webEnvironment = SpringBootTest.WebEnvironment.RANDOM_PORT)
@AutoConfigureMockMvc
class UserController2Test {
    @Test
    void getString(@Autowired MockMvc mvc) throws Exception{
        String url="/pathVariable/user/username1/password1";
        //创建 HttpServletRequest 对象
        MockHttpServletRequestBuilder builder = MockMvcRequestBuilders.get(url);
        //执行请求
        ResultActions action = mvc.perform(builder);
        //获取请求处理状态是否为 200
        action.andExpect(MockMvcResultMatchers.status().isOk());
    }
}
```

运行测试类 UserController2Test，控制台打印如下信息，正常获取到 username 和 password 的值。

```
username=username1,password=password1
```

下面总结@RequestParam、@RequsetBody 和@PathVariable 三个注解的使用场景。@RequestParam 注解用于接收格式为 xx=xx&xx=xx 的数据。可以接收 GET 和 POST 请求数据，POST 请求中 Content-Type 必须为 application/x-www-form-urlencoded，一般用于请求参数较少的场景。@RequsetBody 注解一般用于接收 POST 请求的 JSON 格式数

据,以自定义 Java 对象接收数据,一般用于请求参数较多的场景。@PathVariable 注解利用占位符匹配请求参数,只能接收 GET 请求数据,一般用于请求参数较少的场景。

任务 8.5 Spring MVC 访问动态资源——输出响应

请求处理完毕,需输出响应,Spring MVC 支持使用 Servlet 原生对象——HttpServletResponse 输出响应,并对响应输出做进一步封装,以简化编程。这里主要介绍跳转页面和回写数据这两类最常见的响应。由于目前还未涉及具体的动态页面模板引擎,本节案例以静态页面演示跳转页面功能。

8.5.1 跳转页面

跳转页面是 Web 应用中的基本功能。Spring MVC 默认通过请求转发模式跳转页面,也支持重定向方式跳转页面。下面将分别介绍。

1. 请求转发

请求转发是服务器对同一个请求的二次传递,底层利用 Request 对象的 getRequestDispatcher 方法实现,转发过程中浏览器地址是不变的。Spring MVC 对请求转发进行进一步封装,使用"forward:"前缀实现请求转发跳转页面。当控制器方法中所设置的视图名称以"forward:"为前缀时,Spring MVC 会创建 InternalResourceView 视图,此时的视图名称不会被 Spring MVC 配置文件中所配置的视图解析器解析,而是会将前缀"forward:"去掉,剩余部分作为最终路径转发跳转页面。

Spring MVC 请求转发具体使用如下:假设在 static 目录下有静态页面 index.html,可以使用如下方法进行页面跳转。

```
@RequestMapping(value="/indexForward")
    public String indexForward(){
        return "forward:index.html";        //页面视图资源路径
    }
```

在浏览器中输入地址 http://localhost:8080/indexForward,能够访问到 index.html 页面,请求地址并没有发生变化。

请求转发是 Spring MVC 默认跳转页面使用的模式,使用时 return "forward:index.html"可简写为 return "index.html"。"forward:"前缀后面除了拼接访问的视图资源路径,还能拼接其他控制器方法的 URL,具体示例如下(这种方式也能访问到 index.html 页面):

```
@RequestMapping(value="/indexForwardToUrl")
    public String indexForwardToUrl(){
        return "forward:/forwardIndex";
    }
@RequestMapping(value="/forwardIndex")
    public String forwardIndex(){
        return "forward:index.html";
    }
```

在浏览器中输入地址 http://localhost:8080/indexForwardToUrl，请求会被转发到 forwardIndex 控制器方法处理，跳转到 index.html 页面。

请求转发还可以传递动态数据，因为是同一个请求，数据可直接在当前 Request 域利用 @RequestAttribute 注解获取。Spring MVC 请求转发支持多种方式传递数据，除原生的 HttpServletRequest 对象外，还可利用 ModelAndView/Model 对象传递数据，数据可以是 String、Integer 等简单数据类型，也可以是 Java 对象。ModelAndView 使用示例如下。

```
@RequestMapping(value="/modelAndView")
public ModelAndView modelAndView(@PathVariable String username,
                    ModelAndView modelAndView){
    modelAndView.addObject(属性键,属性值);
    //setViewName 内部也可以为转发的其他请求 URL
    modelAndView.setViewName("视图资源路径");
    return modelAndView;
}
```

该方法返回一个 ModelAndView 对象。ModelAndView 对象内部通过 addObject 方法绑定数据，可为每条数据设定唯一的键，并通过 setViewName 方法设定要跳转的视图页面路径。

ModelAndView 对象将视图和数据绑定在一起，没有实现数据和视图资源的解耦。为解决这个问题，可将 Model 对象和 View 对象拆开，Model 对象只用于传递数据，View 对象只用于跳转页面，前端需借助特定的模板引擎动态整合页面和数据，使数据显示在相应的页面上。Model 对象使用示例如下。

```
@RequestMapping(value="/model")
public String model(@PathVariable String username,Model model){
    model.addAttribute(属性键,属性值);
    return "视图资源路径";           //此处也可以为转发的其他请求 URL
}
```

该方法返回值为 String 类型，为跳转的视图页面路径。Model 对象通过 addAttribute 方法绑定数据，每个数据有唯一的键。

由于 Model 对象传递数据需借助具体的动态模板引擎接收。这里以 ModelAndView 为例演示请求转发过程的数据传递。在 controller 目录下新建一个控制类 UserController3，内部添加如下两个方法，请求转发传递一个 String 类型数据。

```
//请求转发传递数据
@RequestMapping(value="/dispatcher/{username}")
public ModelAndView modelAndView(@PathVariable("username") String username,
ModelAndView modelAndView){
    modelAndView.setViewName("forward:/dispatcher1");
    return modelAndView;
}
@RequestMapping(value="/dispatcher1")
public String model1(@RequestAttribute("username") String username){
```

```
            System.out.println("username="+username);
            return "index.html";
    }
```

上述代码使用 modelAndView 对象将请求转发给映射地址为 model1 的控制器方法处理，model1 控制器方法可通过@RequestAttribute("username") String username 方式获取当前 Request 域的请求数据。

启动项目，在浏览器中输入地址 http://localhost:8080/dispatcher/username，页面跳转到 index.html，控制台打印出如下信息，正常获取了 username 的值。

```
username=username
```

浏览器的开发者工具中可以看到浏览器只发送了一次请求，请求被转发了，如图 8-2 所示。

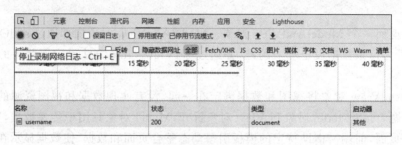

图 8-2　请求转发

2. 请求重定向

请求重定向是服务器利用 Response 对象的 sendRedirect 方法向浏览器发送响应信息，要求浏览器重新发送一个新请求，重定向后请求地址会发送变化。Spring MVC 对请求重定向进行进一步封装，使用"redirect:"前缀实现请求重定向跳转页面，当控制器方法中所设置的视图名称以"redirect:"为前缀时，Spring MVC 会创建 RedirectView 视图，此时的视图名称不会被 Spring MVC 配置文件中所配置的视图解析器解析，而是会将前缀"redirect:"去掉，剩余部分作为最终路径通过重定向的方式实现跳转。

Spring MVC 请求重定向的具体示例如下：假设在 static 目录下有静态页面 index.html，可以使用如下方法进行页面跳转。

```
@RequestMapping(value="/indexRedirct")
    public String indexRedirct(){
        return "redirect:index.html";       //访问的视图资源路径
}
```

在浏览器中输入地址 http://localhost:8080/indexRedirct，能够访问到 index.html 页面，同时浏览器地址改为 http://localhost:8080/index.html。

请求重定向也可以传递动态数据，与请求转发不同，请求重定向是客户端发送了两次请求，因此不能直接通过 Request 域获取请求数据，需手动在请求转发方法中再封装一次数据才能获取。Spring MVC 请求重定向支持多种方式传递数据，除原生的 HttpServletResponse 对象外，还可以利用 ModelAndView/Model 对象和 redirectAttributes 对象传递数据。

（1）ModelAndView/Model 对象。该方式传递的数据是以字符串拼接在请求重定向 URL 的后面，只适用于传递 String、Int 等简单数据类型数据，不支持传递 Java 对象。数据接收可以采用@RequestParam 注解或@ModelAttribute 注解。下面以 ModelAndView 注解为例演示请求重定向过程的数据传递。在控制类 UserController3 中添加如下代码。

```
//请求重定向传递数据
@RequestMapping(value="/redirect/{username}")
public ModelAndView redirect(@PathVariable("username") String username,
    ModelAndView modelAndView){
        //重定向模式下,此处粗体代码不可少,否则就不能正常获取 username 的值
        modelAndView.addObject("username",username);
        modelAndView.setViewName("redirect:/redirect1");
        return modelAndView;
}
@RequestMapping(value="/redirect1")
public String redirect1(@ModelAttribute ("username")String username){
    System.out.println("redirect1 username="+username);
    return "index.html";
}
```

在上述代码中用到了@ModelAttribute 注解，该注解用于获取重定向中 ModelAndView 对象内部封装的数据。接收数据时，@ModelAttribute 注解内部属性值必须和 modelAndView 传递数据的键名保持一致才能正常收到数据。此处也可以采用@RequestParam（"username"）String username 获取数据。

启动项目，在浏览器中输入地址 http://localhost:8080/redirect，页面跳转到 index.html。同时浏览器地址变为 http://localhost:8080/redirect1？username＝username，数据拼接在新的地址 redirect1 后面。控制台打印出如下信息，@ModelAttribute 注解已正常获取到传递的数据。

```
redirect1 username=username
```

在浏览器的开发者工具中可以看到浏览器发送了两次请求，如图 8-3 所示。

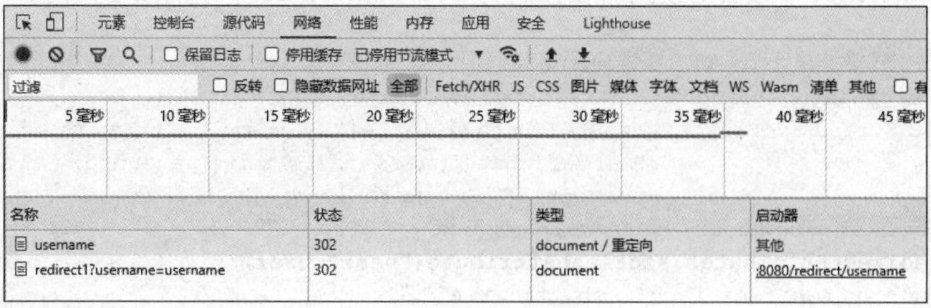

图 8-3 请求重定向

（2）redirectAttributes 对象。除了 ModelAndView 对象外，Spring Boot 还提供 redirectAttributes 对象支持重定向的数据传递。该对象内部有两个方法用于封装数据，分

别是 addAttribute 和 addFlashAttribute。

addAttribute 方法类似于使用 ModelAndView 对象实现重定向，传递的数据以字符串拼接在请求重定向 URL 的后面，只适用于传递 String、Int 等简单数据类型数据，不支持传递 Java 对象。数据接收可以采用@RequestParam 注解或@ModelAttribute 注解。

addAttribute 方法的使用语法如下。

```
redirectAttributes.addAttributie("属性键","属性值");
```

下面演示 addAttribute 方法的使用。在控制类 UserController3 中添加如下代码。

```java
//redirectAttributes 请求重定向传递数据
@RequestMapping("/addAttribute/{username}")
public String addAttribute(@PathVariable("username") String username,
                RedirectAttributes redirectAttributes) {
    redirectAttributes.addAttribute("username",username);
    return "redirect:/addAttribute1";
}
@RequestMapping(value = "/addAttribute1")
public String addAttribute1(@ModelAttribute ("username") String username) {
    System.out.println("username="+username);
    return "index.html";
}
```

启动项目，在浏览器中输入地址 http://localhost:8080/redirect，页面跳转到 index.html。浏览器地址变为 http://localhost:8080/redirect1？username＝username，数据拼接在新的地址 redirect1 后面。控制台打印出如下信息，@ModelAttribute 注解正常获取到传递的数据。

```
username=aaaa
```

addFlashAttribute 方法将数据封装在 Session 中传递，外界不可见，数据一旦在页面获取一次，随即销毁 Session 中的数据。addFlashAttribute 方法比 addAttribute 方法安全性更高，实际应用中更多使用 addFlashAttribute 实现重定向。addFlashAttribute 方法可以传递简单数据类型，也可以传递 Java 对象，数据接收可采用@ModelAttribute 注解。下面演示 addFlashAttribute 方法的使用。在控制类 UserController3 中添加如下代码。

```java
//addFlashAttribute 请求重定向传递数据
@RequestMapping("/addFlashAttribute/{username}/{password}")
public String addFlashAttribute(@PathVariable("username") String username,
                @PathVariable("password") String password,
                RedirectAttributes redirectAttributes) {
    User user=new User(username,password);
    redirectAttributes.addFlashAttribute("user",user);
    return "redirect:/addFlashAttribute1";
}
@RequestMapping(value = "/addFlashAttribute1")
    public String addFlashAttribute1(@ModelAttribute ("user") User user) {
    System.out.println("user="+user);
```

```
        return "index.html";
}
```

启动项目,在浏览器中输入地址。

```
http://localhost:8080/addFlashAttribute/username/password
```

页面跳转到 index.html。浏览器地址变为 http://localhost:8080/addFlashAttribute1;jsessionid=68B8D8FC6E52FF714EE4A158907EB8CF。

在浏览器地址中看见了当前的 sessionid,证明 addFlashAttribute 方法利用 Session 对象传递参数。控制台打印出如下信息,@ModelAttribute 注解正常获取到传递的数据。

```
user=User(username=username, password=password)
```

下面介绍一种更加通用的传值方式,无论是请求转发还是重定向,都在同一个 Session 对象中。因此都可以利用 Session 对象传递数据。以请求重定向为例具体使用如下。在控制类 UserController3 中添加如下代码。

```
@RequestMapping("/setSession/{username}/{password}")
public String setSession(@PathVariable("username") String username,
                @PathVariable("password") String password,
                HttpSession httpSession) {
    User user=new User(username,password);
    httpSession.setAttribute("user",user);
    return "redirect:/getSession";
}
@RequestMapping(value = "/getSession")
public String getSession(HttpSession httpSession) {
    System.out.println("user="+(User)httpSession.getAttribute("user"));
    return "index.html";
}
```

在浏览器中访问地址 http://localhost:8080/setSession/username/password,页面跳转到 index.html。控制台打印如下信息 user = User(username = username, password = password)。浏览器地址变为 http://localhost:8080/getSession;jsessionid = 15DA6B9-DAFBA8311FDA9218CEFAC1D34。

获取的数据和 addFlashAttribute 方法效果一样,再次证明 addFlashAttribute 方法底层采用 session 传递数据。

8.5.2 回写数据

如果请求处理完毕,不需跳转页面而只将数据写回当前 URL,就可采用@ResponseBody 注解。@ResponseBody 注解一般用于前端 AJAX 异步请求数据。在微服务架构中,对外提供服务的 Web API 接口也大多以这种类型存在。

当请求处理方法被@RequestMapping 注解和@ResponseBody 注解同时修饰时,方法返回值不会被解析为跳转路径,而是统一被解析为 JSON 格式字符串,写入 Response 消息体中,不进行页面跳转。@ResponseBody 注解使用示例如下。

```
@RequestMapping(value="/returnToNow/{username}/{password}")
@ResponseBody
public User returnToNow(@PathVariable("username") String username,
@PathVariable("password") String password){
    User user=new User(username,password);
    return user;
}
```

该方法返回 User 对象,User 对象被@ResponseBody 注解解析成 JSON 字符串并输出到当前 URL 对应的页面。在浏览器中输入地址 http://localhost:8080/returnToNow/username/password,当前页面显示如下字符串,数据成功回写到当前页面。

```
{"username":"username","password":"password"}
```

为了方便,此处还可以直接在控制类上使用@RestController 注解替换@Controller 注解,这样该控制器类内部的所有请求处理方法不再需要添加@ResponseBody 注解,@RestController 注解使控制器内部所有请求处理方法直接向当前 URL 输出 JSON 格式响应数据,而不跳转页面。

任务 8.6 Spring MVC Restful 风格编程

当前的 Web API 接口大多采用 Restful 风格实现,Restful 具有结构清晰、符合标准、易于理解以及扩展方便等特点,受到越来越多互联网公司的青睐。本任务主要介绍 Spring MVC 如何实现 Restful 风格的 Web API 接口。

8.6.1 初识 Restful 风格

Restful 也称为 REST,是英文 representational state transfer 的简称。Restful 是一种软件架构风格或设计风格。在 Restful 风格中,每个访问路径代表一种资源,客户端对服务端的请求就是对服务端各种资源的操作,该操作主要分为增、删、改、查四类,分别使用 POST、DELETE、PUT 和 GET 请求与之相对应,客户端可根据需求使用不同的请求访问服务端。如需传递参数,简单类型参数可以放在请求路径中传递,Java 对象可以封装在请求体中传递。

如传统的增、删、改、查 URL 地址如下。

```
http://localhost/user/query/1        GET 请求,通过 id 查询用户
http://localhost/user/save           POST 请求,新增用户
http://localhost/user/update         POST 请求,修改用户
http://localhost/user/delete/1       GET 请求,通过 id 删除用户
```

使用 Restful 风格后,URL 就变成了下面的格式。

http://localhost/user/{id}	GET 请求,通过用户 id 查询用户
http://localhost/user	POST 请求,新增用户
http://localhost/user	PUT 请求,修改用户
http://localhost/user/{id}	DELETE 通过 id 删除用户

从上述案例可以看出,采用 Restful 风格后,具体的动作封装在请求类型中,访问路径中不再带有 query、save、update 和 delete 等动词,而只能有名词,这个名词就是访问的资源,通常是数据库中某数据表的名称。在数据传递方面,简单类型的数据可以直接通过斜杠拼接在 URL 后面,Java 对象以 JSON 字符串形式封装在 POST 和 PUT 请求体中传递。Restful 风格接口响应的数据也统一以 JSON 格式字符串返回客户端。

Restful 风格使 Web API 接口命名更加规范化和统一化,同时将具体的动作封装在请求类型中,使外界无法从 URL 直接得知客户端对服务端做了何种操作,增强了安全性。

Restful 风格简介

8.6.2 Spring MVC 实现 Restful 风格编程

Spring MVC 可利用@RestController 注解、@PathVariable 注解和@RequsetBody 注解实现 Restful 风格增、删、改、查的 Web API 接口。使用时须注意以下事项。

(1) @RestController 注解:该注解修饰在控制器类上,控制器内部所有请求处理方法不跳转页面,直接在当前 URL 返回数据。

(2) @PathVariable 注解:该注解用于接收简单数据类型参数,一般用在查询和删除接口上获取数据。

(3) @RequsetBody 注解:该注解用于接收请求体中 JSON 字符串格式的 Java 对象类型参数,一般用在新增和修改接口上获取数据。

(4) @RestController 注解:可使用@Controller 注解和@ResponseBody 注解组合代替。

下面编程实现 Restful 风格的增、删、改、查 Web API 接口。在 controller 目录下新建 RestfulController 类,内部定义增、删、改、查 4 个 Restful 风格接口。

【RestfulController.java】

```
package springdemo8.controller;
import org.springframework.web.bind.annotation.*;
import springdemo8.domain.User;
//如果此处为@Controller 注解,内部每个方法都必须要加@ResponseBody 注解
@RestController
public class RestfulController {
    //模拟通过用户名查询用户
    @GetMapping("/restUser/{username}")
```

```java
    public User getUser(@PathVariable("username") String username){
        //此处省略数据库查询用户代码,模拟一个用户输出
        User user=new User(username,"password");
        return user;
    }
    //模拟添加用户
    @PostMapping("/restUser")
    public User addUser(@RequestBody User user){
        //此处省略数据库新增用户代码,直接输出收到的 User 对象
        return user;
    }
    //模拟修改用户
    @PutMapping("/restUser")
    public User updateUser(@RequestBody User user){
        //此处省略数据库修改用户代码,直接输出收到的 User 对象
        return user;
    }
    //模拟通过用户名删除用户
    @DeleteMapping("/restUser/{username}")
    public String deleteUser(@PathVariable("username") String username){
        //此处省略数据库删除用户代码,直接输出收到的 username
        return username;
    }
}
```

上述代码中利用@RestController 注解修饰整个类,内部分别利用@GetMapping、@PostMapping、@PutMapping 和@DeleteMapping 这 4 个注解映射查、增、改和删操作。其中,@GetMapping 注解和@DeleteMapping 注解的请求参数利用@PathVariable 注解获取,@PostMapping 注解和@PutMapping 注解的请求参数利用@RequestBody 注解获取。

创建测试类 RestfulControllerTest.java,内部利用 MockMvc 工具测试 4 个接口。这里试着传递中文数据,并设置响应编码格式为 UTF8,以防止响应信息中文乱码。

【RestfulControllerTest.java】

```java
package springdemo8.controller;
import com.fasterxml.jackson.databind.ObjectMapper;
import org.junit.jupiter.api.Test;
import org.springframework.beans.factory.annotation.Autowired;
import org.springframework.boot.test.autoconfigure.web.servlet.AutoConfigureMockMvc;
import org.springframework.boot.test.context.SpringBootTest;
import org.springframework.http.MediaType;
import org.springframework.test.web.servlet.MockMvc;
import org.springframework.test.web.servlet.ResultActions;
import org.springframework.test.web.servlet.request.MockHttpServletRequestBuilder;
import org.springframework.test.web.servlet.request.MockMvcRequestBuilders;
import org.springframework.test.web.servlet.result.MockMvcResultMatchers;
import springdemo8.domain.User;
@SpringBootTest(webEnvironment = SpringBootTest.WebEnvironment.RANDOM_PORT)
@AutoConfigureMockMvc
```

```java
class RestfulControllerTest {
    @Test
    void sendDataMain(@Autowired MockMvc mvc) throws Exception{
        sendData(mvc,"get","/restUser");          //查询用户
        sendData(mvc,"delete","/restUser");       //删除用户
        sendData(mvc,"put","/restUser");          //修改用户
        sendData(mvc,"post","/restUser");         //新增用户
    }
    void sendData(MockMvc mvc,String type,String url)throws Exception{
        MockHttpServletRequestBuilder builder;
        ObjectMapper objectMapper=new ObjectMapper();
        User user=new User("用户 1","密码 1");
        if(type.equals("get")){
            builder = MockMvcRequestBuilders.get(url+"/用户 1");
        }
        else if(type.equals("delete")){
            builder = MockMvcRequestBuilders.delete(url+"/用户 1");
        }
        else if(type.equals("put")){
            builder = MockMvcRequestBuilders.put(url)
                    .contentType(MediaType.APPLICATION_JSON_VALUE)
                    .content(objectMapper.writeValueAsString(user));
        }
        else{
            builder = MockMvcRequestBuilders.post(url)
                    .contentType(MediaType.APPLICATION_JSON_VALUE)
                    .content(objectMapper.writeValueAsString(user));
        }
        //执行请求
        ResultActions action = mvc.perform(builder);
        //获取请求处理状态是否为 200
        action.andExpect(MockMvcResultMatchers.status().isOk());
        //获取响应信息,设置编码为 UTF8,否则中文响应乱码
        MockHttpServletResponse response = action.andReturn().getResponse();
        response.setCharacterEncoding("UTF-8");
        System.out.println(type+" result="+response.getContentAsString());
    }
}
```

运行测试类,控制台打印 4 个接口的返回值如下。

```
get result={"username":"用户 1","password":"password"}
delete result=用户 1
put result={"username":"用户 1","password":"密码 1"}
post result={"username":"用户 1","password":"密码 1"}
```

任务 8.7　Spring MVC 拦截器

在前面的内容中,我们在任何情况下只要输入地址,都能访问相应的控制器方法获取服务端资源,这是很不安全的。一个成熟的项目都需要对内部资源的访问做一些控制措施,例如某些资源游客可以访问,某些资源登录之后才可以访问,某些资源需具备特定的权限才可以访问。实现这类功能就需要用到拦截器。拦截器是 Spring MVC 中的一个扩展功能,能够拦截所有的资源访问并进行逻辑控制,同时拦截器也能够很方便地在 Spring Boot 项目中使用。本任务将介绍如何在 Spring Boot 项目中使用 Spring MVC 拦截器。

8.7.1　定义拦截器

拦截器是 Web 开发中的常用功能,用于增强客户端对服务端资源的访问控制。在 Spring MVC 中,拦截器的底层是利用 Spring AOP 思想对所有进入控制器方法的请求进行增强处理。

拦截器的定义非常简单,一个 Java 类只要实现 HandlerInterceptor 接口,该类就是一个拦截器类。HandlerInterceptor 接口中定义以下三个方法,分别对应于不同的拦截时间。

(1) preHandle 方法:preHandle 方法在客户端请求进入控制器请求处理方法之前执行。方法返回值为 Boolean 类型,如返回值为 true,则放行请求;如返回值为 false,则拦截请求不放行。一般用于权限认证,判断用户是否登录,统一设置请求编码格式等。

(2) postHandle 方法:postHandle 方法在控制器请求处理方法执行过程中执行,即请求处理方法已执行但还未解析视图。方法返回值为 void 类型。一般用于统一设置响应编码格式等。

(3) afterCompletion 方法:afterCompletion 方法会在控制器请求处理方法全部执行完后执行,即视图渲染结束之后执行。方法返回值为 void 类型。一般用于资源清理、记录日志信息等。

在代码中定义拦截器只需创建一个普通的 Java 类,该类实现 HandlerInterceptor 接口,并根据需要选择实现 HandlerInterceptor 接口的部分或全部方法即可。

例如在 Web 开发中,我们经常会拦截用户请求,如果用户登录过就放行,否则跳转回登录页面。可使用如下代码实现拦截器。

```java
public class LoginInterceptor implements HandlerInterceptor {
    @Override
    public boolean preHandle(HttpServletRequest request, HttpServletResponse response,
                    Object handler) throws Exception {
        String userinfo=request.getSession().getAttribute("user").toString();
        if(null==userinfo){
            response.sendRedirect("/login.html");
            return false;
        }else{
            return true;
```

```
        }
    }
}
```

上述代码拦截用户请求,并从 Session 对象中获取用户信息。如果用户未登录过系统,则 Session 对象的信息为空,拦截器拒绝用户访问跳转回登录页面,否则放行请求。

8.7.2 使用拦截器

当定义好一个拦截器后,该如何使用呢?在 Spring Boot 项目中使用拦截器非常简单,只需自定义一个配置类并实现 WebMvcConfigurer 接口。WebMvcConfigurer 接口是扩展 Spring MVC 功能的一个接口,内部定义了很多扩展方法,用户只需覆写对应方法即可扩展 Spring MVC 功能。使用拦截器需在配置类内部覆写 WebMvcConfigurer 接口的 addInterceptors 方法来注册已定义的拦截器。注册完毕,指定相应的拦截规则,设置拦截哪些请求,放行哪些请求。addInterceptors 方法内的 InterceptorRegistry 对象可注册一个或多个拦截器,如注册多个拦截器有先后执行关系,可通过 InterceptorRegistry 对象的 order 方法设置拦截器的执行顺序。

下面通过一个案例演示拦截器的使用。拦截器功能为根据时间对客户端资源访问进行拦截控制,客户端只能在某时间段内访问服务端特定 URL 的接口资源,如访问其他资源则拒绝访问跳转 404 页面,如超出访问时间段也拒绝访问跳转 404 页面。

在 springdemo8 目录下新建 interceptor 目录,在 interceptor 目录下新建两个拦截器 TimeInterceptor 和 URLInterceptor,拦截器 TimeInterceptor 用于设定用户访问时间段为每天 20:00—23:00,其余时间拒绝访问,跳转到 404_time.html 页面;拦截器 URLInterceptor 用于设定用户只能访问 URL 为 indexforward 的接口,其他接口拒绝访问,跳转到 404_url.html 页面。

【TimeInterceptor.java】

```
package springdemo8.interceptor;
import jakarta.servlet.http.HttpServletRequest;
import jakarta.servlet.http.HttpServletResponse;
import org.springframework.context.annotation.Configuration;
import org.springframework.web.servlet.HandlerInterceptor;
import java.time.LocalTime;
@Configuration
public class TimeInterceptor implements HandlerInterceptor {
    @Override
    public boolean preHandle(HttpServletRequest request, HttpServletResponse response,
                    Object handler) throws Exception {
        LocalTime begintime=LocalTime.of(20,0);
        LocalTime endttime=LocalTime.of(23,0);
        LocalTime localTime=LocalTime.now();
        //只允许在每天晚上 20:00—23:00 访问,否则重定向跳转 404_time.html
        if(localTime.isAfter(begintime)&&localTime.isBefore(endttime)){
            return true;
        }else{
```

```
            response.sendRedirect("/404_time.html");
            return false;
        }
    }
}
```

【URLInterceptor.java】

```
package springdemo8.interceptor;
import jakarta.servlet.http.HttpServletRequest;
import jakarta.servlet.http.HttpServletResponse;
import org.springframework.context.annotation.Configuration;
import org.springframework.web.servlet.HandlerInterceptor;
@Configuration
public class URLInterceptor implements HandlerInterceptor {
    @Override
    public boolean preHandle(HttpServletRequest request, HttpServletResponse response,
                    Object handler) throws Exception {
        String url=request.getRequestURI();
        //只能访问 URL 为 indexforward 的接口,否则重定向跳转到 404_url.html
        if(url.equals("indexforward")){
            return true;
        }else{
            response.sendRedirect("/404_url.html");
            return false;
        }
    }
}
```

在 springdemo8 目录下新建 config 目录,在 config 目录下新建 InterceptorConfig 类,在 InterceptorConfig 类实现 WebMvcConfigurer 接口,内部覆写 addInterceptors 方法;在 addInterceptors 方法内部依次注册两个拦截器并设定拦截规则。

【InterceptorConfig.java】

```
package springdemo8.config;
import org.springframework.beans.factory.annotation.Autowired;
import org.springframework.context.annotation.Configuration;
import org.springframework.web.servlet.config.annotation.InterceptorRegistry;
import org.springframework.web.servlet.config.annotation.WebMvcConfigurer;
import springdemo8.interceptor.TimeInterceptor;
import springdemo8.interceptor.URLInterceptor;
@Configuration
public class InterceptorConfig implements WebMvcConfigurer {
    @Autowired
    private TimeInterceptor timeInterceptor;
    @Autowired
    private URLInterceptor urlInterceptor;
    @Override
```

```
    public void addInterceptors(InterceptorRegistry registry) {
        registry.addInterceptor(timeInterceptor)
                //设定拦截哪些请求
                .addPathPatterns("/**")
                //设定不拦截哪些请求,一般是静态资源CSS、JS、图片等
                .excludePathPatterns("/","/index.html",
                    "/404_time.html","/404_url.html")
                //设定多个拦截器的执行顺序,order内部数字越小则优先级越高
                .order(1);
        registry.addInterceptor(urlInterceptor)
                .addPathPatterns("/**")
                .excludePathPatterns("/","/index.html",
                    "/404_time.html","/404_url.html")
                .order(2);
    }
}
```

其中,addPathPatterns("/**")方法用于拦截所有请求;excludePathPatterns方法一般用于排除一些不需拦截的静态资源访问,这里设定排除根目录、index.html、404_time.html和404_url.html页面。order方法用于设定多个拦截器的执行顺序,order内部数字越小则优先级越高,这里设定先按照访问时间拦截,如果访问时间不符合要求,拒绝访问;访问时间符合要求,再进行URL校验。

在resources的static目录下新建两个访问拒绝页面404_time.html和404_url.html,内部分别填写文本not allowed time和not allowed url。

启动Springdemo8Application,23:25在浏览器中输入http://localhost:8080/indexforward,访问被拒绝,页面跳转到404_time.html页面,显示文字not allowed time;如果在晚上20:00—23:00访问indexforward,访问成功,跳转到index.html页面,显示文字hello world;如果在20:00—23:00访问其他URl,则跳转到404_url.html页面,显示文字not allowed url。

8.7.3 拦截器和过滤器

与拦截器功能类似的还有过滤器,过滤器和拦截器都能够实现权限控制和日志记录等功能。但是Spring框架优先推荐使用拦截器,具体原因如下。

(1) 过滤器是Servlet中的概念,只能用于Web应用,但是拦截器可以用于普通的Java应用。

(2) 拦截器是SpringMVC的扩展功能,是Spring框架的一部分。Spring的任何资源和对象都可以直接注入拦截器中使用,如Bean对象,数据源,事务管理等,而过滤器却不行。

(3) 过滤器只能在Servlet的前后起作用,而拦截器能在控制器方法前后执行,使用更加灵活,粒度更小。

因此在Spring项目中,一般优先使用拦截器实现相关功能。

任务 8.8　Spring MVC 文件上传和下载

文件上传与下载是 Web 应用中的常见功能，SpringMVC 对此也提供相应支持。本任务将介绍如何在 Spring Boot 项目中利用 Spring MVC 上传和下载文件。

8.8.1　Spring MVC 文件上传

SpringMVC 利用@RequestPart 注解接收上传的文件，上传的文件以 MultipartFile 对象类型存在，并利用 MultipartFile 对象的 transferTo 方法将文件写入特定的目录，可以上传一个文件，也可同时上传多个文件。单个文件上传功能的接口示例如下。

```java
@PostMapping("/upload")
public String upload(@RequestPart("file") MultipartFile file) throws IOException {
    file.transferTo(new File("存储上传文件的目录" + "文件名称"))
}
```

如上传多个文件，可以使用 MultipartFile 对象集合接收，然后遍历 MultipartFile 对象集合，对其中的每个文件依次执行 transferTo 方法。多个文件上传功能的接口示例如下。

```java
@PostMapping("/upload")
public String upload(@RequestPart("files") List<MultipartFile> files) throws IOException {
    for (MultipartFile multipartFile : files) {
        multipartFile.transferTo(new File("存储上传文件的目录" + "文件名称"))
    }
}
```

下面编程实现上传单个和多个文件，并存储在 D 盘的 springdemo8_uploadfile 目录下。这里先将上个任务中定义的拦截器配置类 InterceptorConfig 上的@Configuration 注释掉，以免文件上传请求被拦截。在 SpringDemo8 项目的配置文件 application.yaml 中添加如下配置，配置文件上传目录为 D 盘的 springdemo8_uploadfile 目录。

```
uploadfilepath: D:/springdemo8_uploadfile/
```

在 controller 目录下新建 FileController 类，内部定义 uploadonefile 和 uploadmanyfile 两个方法，分别用于单文件上传和多文件上传。

【FileController.java】

```java
package springdemo8.controller;
import org.springframework.beans.factory.annotation.Value;
import org.springframework.stereotype.Controller;
import org.springframework.web.bind.annotation.PostMapping;
```

```java
import org.springframework.web.bind.annotation.RequestPart;
import org.springframework.web.bind.annotation.ResponseBody;
import org.springframework.web.multipart.MultipartFile;
import java.io.File;
import java.io.IOException;
import java.util.List;
@Controller
public class FileController {
    //注入配置文件中定义的文件存储目录
    @Value("${uploadfilepath}")
    private String filePath;
    //上传单个文件
    @PostMapping("/uploadOneFile")
    @ResponseBody
    public String uploadOneFile(@RequestPart("file") MultipartFile file)
            throws IOException {
        //判断文件存储目录是否存在,不存在则创建目录
        File directory = new File(filePath);
        if (!directory.exists()) {
            directory.mkdirs();
        }
        //判断文件是否为空,不为空则保存文件
        if (!file.isEmpty()) {
            saveFile(file);
            return "success";
        }else{
            return "file is empty";
        }
    }
    //上传多个文件
    @PostMapping("/uploadManyFile")
    @ResponseBody
    public String uploadManyFile(@RequestPart("files") List<MultipartFile> files)
            throws IOException {
        if (files.size() > 0) {
            for (MultipartFile multipartFile : files) {
                if (!multipartFile.isEmpty()) {
                    saveFile(multipartFile);
                }
            }
            return "success";
        }else{
            return "filelist is empty";
        }
    }
    //保存文件到指定目录
    public void saveFile(MultipartFile file) throws IOException {
        //获取文件名
        String name = file.getOriginalFilename();
        //将文件写到指定目录
```

```
            file.transferTo(new File(filePath + name));
    }
}
```

创建测试类 FileControllerTest,内部利用 MockMVC 工具测试单文件上传和多文件上传接口。

【FileControllerTest.java】

```
package springdemo8.controller;
import org.junit.jupiter.api.Test;
import org.springframework.beans.factory.annotation.Autowired;
import org.springframework.boot.test.autoconfigure.web.servlet.AutoConfigureMockMvc;
import org.springframework.boot.test.context.SpringBootTest;
import org.springframework.http.MediaType;
import org.springframework.mock.web.MockMultipartFile;
import org.springframework.test.web.servlet.MockMvc;
import org.springframework.test.web.servlet.request.MockMvcRequestBuilders;
import org.springframework.test.web.servlet.result.MockMvcResultMatchers;
import java.io.File;
import java.io.FileInputStream;
@SpringBootTest(webEnvironment = SpringBootTest.WebEnvironment.RANDOM_PORT)
@AutoConfigureMockMvc
class FileControllerTest {
    @Test                                   //单文件上传
    void uploadOneFile(@Autowired MockMvc mvc) throws Exception {
        File file = new File("D:\\1.pdf");
        MockMultipartFile firstFile =
                new MockMultipartFile("file", "1.pdf",
                    MediaType.TEXT_PLAIN_VALUE, new FileInputStream(file));
        mvc.perform(MockMvcRequestBuilders.multipart("/uploadonefile")
                .file(firstFile)
                .contentType(MediaType.MULTIPART_FORM_DATA_VALUE))
                .andExpect(MockMvcResultMatchers.status().isOk());
    }
    @Test                                   //多文件上传
    void uploadManyFile(@Autowired MockMvc mvc) throws Exception {
        File file1 = new File("D:\\2.docx");
        File file2 = new File("D:\\3.xls");
        MockMultipartFile firstFile = new MockMultipartFile("files", "2.docx",
                MediaType.TEXT_PLAIN_VALUE, new FileInputStream(file1));
        MockMultipartFile secondFile = new MockMultipartFile("files", "3.xls",
                MediaType.TEXT_PLAIN_VALUE, new FileInputStream(file2));
        mvc.perform(MockMvcRequestBuilders.multipart("/uploadmanyfile")
                .file(firstFile).file(secondFile)
                .contentType(MediaType.MULTIPART_FORM_DATA_VALUE))
                .andExpect(MockMvcResultMatchers.status().isOk());
    }
}
```

在 D 盘根目录下新建 1.pdf、2.docx 和 3.xls 三个文件，依次运行测试类的 uploadOneFile 和 uploadManyFile 方法，待程序运行完毕方法，D 盘下自动创建了 springdemo8_uploadfile 目录，该目录下有 1.pdf、2.docx 和 3.xls 三个文件，如图 8-4 所示。证明单文件上传和多文件上传都执行成功。

名称	修改日期	类型	大小
1.pdf	2023/5/15 17:48	WPS PDF 文档	440 KB
2.docx	2023/5/15 17:48	DOCX 文档	58 KB
3.xls	2023/5/15 17:48	XLS 工作表	63 KB

图 8-4 文件上传结果

Spring 默认上传文件大小客户端为 10MB，服务端为 1MB。如需上传较大文件，可修改 application.yaml 配置文件中 Spring 默认上传文件大小。如下面代码将服务端文件大小和客户端请求文件大小的上限都修改为 100MB。

```yaml
spring:
  servlet:
    multipart:
      max-file-size: 100MB #服务端文件大小限制，默认为 1MB
      max-request-size: 100MB #客户端请求文件大小限制，默认为 10MB
```

8.8.2　Spring MVC 文件下载

Spring MVC 文件下载可直接利用 HttpServletResponse 对象实现。只需设置响应头为"Content-Disposition：attachment;filename=下载的文件名"，当在浏览器发送文件下载请求时，文件会以附件形式从浏览器下载。下面编程实现浏览器下载已上传的文件 1.pdf。在 FileController 类中添加如下文件下载的接口，下载的文件名以 Restful 风格拼接在下载请求 URL 中。

```java
//文件下载
@GetMapping("/downloadFile/{filename}")
public void downloadFile(@PathVariable("filename") String filename,
                    HttpServletResponse response) throws Exception{
    //根据文件名找到 filePath 目录下的文件
    File file=new File(filePath+filename);
    //设置响应头，以附件的形式下载
    //文件名一定要为 UTF-8 编码，否则中文文件名会出现乱码
    response.setHeader("content-disposition",
           "attachment;filename="+ URLEncoder.encode(filename,"UTF-8"));
    //复制文件，通过 response 以流的形式输出
    IOUtils.copy(new FileInputStream(file),response.getOutputStream());
}
```

运行 Springdemo8Application 启动类，在浏览器中输入地址。

http://localhost:8080/downloadFile/1.pdf

文件名为1.pdf的文件就会被下载下来。

任务8.9　综合案例：员工信息管理

在Web应用的MVC三层架构中，Spring MVC实际上是MVC三层架构中的Controller层，又称为Web API接口层。Controller层内部定义请求处理方法接收并处理View层的访问资源请求，请求处理方法内部调用Model层接口去访问资源，获取资源后对View层页面输出。在Web API接口的编码中，为实现代码解耦，Model层又细分为业务层和持久层，业务层专注于业务逻辑的封装，持久层专注于访问数据资源。这里以一个综合案例——员工信息管理为例，编写对员工信息的增、删、改、查操作（涉及控制层、业务层和持久层代码），实现基于Restful风格的增、删、改、查Web API接口。

8.9.1　案例任务

在SpringDemo8项目的基础上，基于MySQL数据库中的数据库表employee并利用SpringMVC+Mybatis编程实现基于Restful风格的增、删、改、查Web API接口。employee表字段为(emp_id,emp_name,emp_gender,emp_dept和emp_birth)。

8.9.2　案例分析

该案例实现基于Restful风格的增、删、改、查Web API接口。涉及控制层、业务层和持久层代码的编写和调用，是一个结合了项目5的Mybatis数据操作和项目8的Spring MVC编程的综合案例。案例实现步骤如下：

(1) 在当前项目中引入MySQL、Mybatis依赖并在application.yaml文件中进行相关配置。
(2) 编写数据持久层代码，实现employee数据表的增、删、改、查操作方法。
(3) 编写业务层代码，调用数据持久层增、删、改、查方法，实现增、删、改、查业务接口。
(4) 编写控制器层请求处理方法，调用增、删、改、查业务接口，实现Restful风格的增、删、改、查Web API接口。

8.9.3　任务实施

由于篇幅有限，任务实施具体步骤采用二维码形式展示。

综合案例的实施步骤

小　　结

本项目详细介绍了如何利用 Spring Boot 集成 Spring MVC 开发 Web 应用的后端，包括请求响应的处理、Restful 风格 API 的编程实现、拦截器的使用、文件的上传和下载等。最后以一个综合案例——员工信息管理，演示如何利用 Spring MVC 整合 Mybatis 开发完整的基于 Restful 风格的增、删、改、查 Web API 接口。

课后练习：学生信息管理

模仿本项目的任务 8.9，在 MySQL 数据库中创建数据库表 student，student 表字段为 (stu_id, stu_name, stu_gender, stu_class 和 stu_birth)，利用 Spring MVC+Mybatis 编程实现基于 Restful 风格的增、删、改、查 Web API 接口。

项目 9　Spring Boot Web 应用开发——前端

在项目 8 中我们学习了在 Spring Boot 项目中引入 Spring MVC 框架实现 Web 后端。本项目将介绍 Spring Boot Web 应用开发的前端内容，即前端如何和 Spring MVC 后端实现数据交互。

任务 9.1　了解 Spring Boot Web 应用前端实现方式

Spring Boot Web 应用的前端有两种实现方式：一种是前后端不分离，即前后端代码统一写在一个 Spring Boot 项目之中；另一种是前后端分离，后端是单独的 Spring Boot 项目，前端也是一个单独的前端项目。

1. 前后端不分离

如果采用的是前后端不分离的方式，在 Spring Boot 中需要使用一些主流的后端模板引擎来渲染数据。模板引擎实现了数据渲染时用户界面与业务数据的分离。在 Spring Boot 项目中，数据通过 Spring MVC 中的 ModelAndView 对象传递 Model 数据到前端页面，前端页面使用相应的后端模板引擎对数据进行解析，使用自身的语法规则对数据进行渲染。目前常用的后端模板引擎有 JSP 和 Thymeleaf。其中 JSP 是早期 Web 开发中应用最多的模板引擎，目前还仍有不少老项目在使用，Spring Boot 项目兼容 JSP 模板引擎，但并不推荐使用。Spring Boot 项目默认推荐使用 Thymeleaf 模板引擎，并提供了默认配置。因此，在前后端不分离的方式下，在 Spring Boot 项目中应尽量使用 Thymeleaf 模板引擎。

2. 前后端分离

如果采用的是前后端分离的方式，Spring Boot 项目就完全变成了一个服务端，内部没有任何前端资源，仅对外提供各种 Web API 接口，而所有的前端页面和资源都构建在另一个独立的前端项目中。前端项目访问 Spring Boot 项目的各个 Web API 获取数据并渲染页面，同时自身也负责前端页面的跳转。前后端分离的方式下构建前端项目的框架有 Vue、React 和 Angular 等。其中 Vue 因其使用简单且易上手而成为前端初学者使用最多的框架。

后续内容将以任务 8.9 员工信息管理为例介绍如何分别利用 JSP、Thymeleaf 和 Vue 实现 Spring Boot Web 应用的前端数据渲染。其中持久层和业务层代码不变，Controller 层根据需要进行修改，增加页面的跳转功能。

任务 9.2 利用 JSP 模板引擎实现前端功能

JSP 是早期 Web 开发中应用最多的模板引擎,功能强大,目前还仍有不少老项目在使用。Spring Boot 项目兼容但不推荐使用 JSP 做模板引擎,只有一些老项目还会需要使用 JSP 做模板引擎。

9.2.1 初识 JSP 模板引擎

JSP 模板引擎在前端页面中可通过 Java 代码或 EL 表达式获取后端传递的数据替换掉模板上的静态数据。在前端页面直接嵌入 Java 代码会降低代码的可读性,不利于维护,因此实际应用中多使用 EL 表达式获取后端传递数据,如有一些逻辑判断可借助 JSP 标签实现。一般服务端传递数据大概有如下三类。

```
model.addAttribute("name",username);           //发送简单类型数据
model.addAttribute("user",user);               //发送 Java 对象数据
model.addAttribute("userlist",userlist);       //发送 Java 对象集合数据
```

页面可以使用 EL 表达式和 JSP 标签接收数据。

获取简单数据可以使用 ${name};获取 Java 对象某属性可以使用 ${user.username};获取集合数据可以使用<c:forEach var="user" items="${userlist}"></c:forEach>。

如页面要向服务端请求传递数据,必须保证页面所传递的数据有 name 属性,否则控制器无法通过 name 获取数据。如页面如下:

```
<p>用户名:<input name="username" type="text"></p>
<p>密码:<input name="password" type="text"></p>
```

控制器才能够通过(String username,String password)获取请求数据。

9.2.2 Spring Boot 引入并配置 JSP 模板引擎

这里以项目 8 任务 8.9 员工信息管理为基础建立一个新的 Spring Boot 项目 SpringDemo9_Jsp,在 SpringDemo9_Jsp 项目的 pom.xml 文件下引入 JSP 相关依赖和数据库相关依赖。其中<jstl>标签依赖必须引入以 jakarta 开头的依赖而不是 javax。

```xml
<dependencies>
    <dependency>
        <groupId>org.springframework.boot</groupId>
        <artifactId>spring-boot-starter-web</artifactId>
    </dependency>
    <dependency>
        <groupId>org.projectlombok</groupId>
        <artifactId>lombok</artifactId>
        <optional>true</optional>
    </dependency>
```

```xml
    <!-- 引入 JSP 解析器-->
    <dependency>
        <groupId>org.apache.tomcat.embed</groupId>
        <artifactId>tomcat-embed-jasper</artifactId>
        <scope>compile</scope>
    </dependency>
    <!--JSP 页面使用 JSTL 标签-->
    <dependency>
        <groupId>org.glassfish.web</groupId>
        <artifactId>jakarta.servlet.jsp.jstl</artifactId>
        <version>3.0.1</version>
    </dependency>
    <dependency>
        <groupId>jakarta.servlet.jsp.jstl</groupId>
        <artifactId>jakarta.servlet.jsp.jstl-api</artifactId>
        <version>3.0.0</version>
    </dependency>
    <dependency>
        <groupId>taglibs</groupId>
        <artifactId>standard</artifactId>
        <version>1.1.2</version>
    </dependency>
    <!-- 引入 Mybatis、MySQL 驱动和 Druid-->
    <dependency>
        <groupId>org.mybatis.spring.boot</groupId>
        <artifactId>mybatis-spring-boot-starter</artifactId>
        <version>3.0.1</version>
    </dependency>
    <dependency>
        <groupId>com.mysql</groupId>
        <artifactId>mysql-connector-j</artifactId>
    </dependency>
    <dependency>
        <groupId>com.alibaba</groupId>
        <artifactId>druid-spring-boot-starter</artifactId>
        <version>1.2.16</version>
    </dependency>
    <dependency>
        <groupId>org.springframework.boot</groupId>
        <artifactId>spring-boot-starter-test</artifactId>
        <scope>test</scope>
    </dependency>
</dependencies>
```

将项目 8 中项目的 Mapper 目录、service 目录和 domain 目录下的 Employee 实体类直接复制过来，重用原来的代码。

SpringBoot 项目要使用 JSP 模板，一般会把所有 JSP 文件放到 webapp 目录下面，这时需要手动创建 webapp 目录，并配置项目识别 webapp 目录为 Web 资源访问目录，如图 9-1 所示。本项目 JSP 放置路径为 webapp/WEB-INF/jsp/。

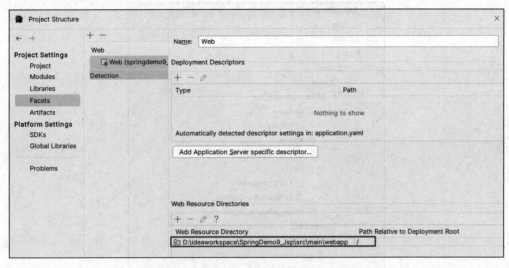

图 9-1 设置 Web 资源访问目录为 webapp

在项目的 application.yaml 配置文件中配置数据库相关信息和 JSP 视图解析器返回文件的前后缀。

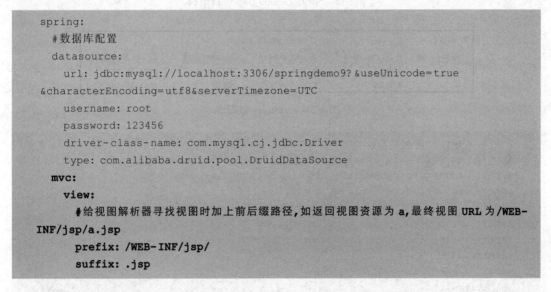

最终 SpringDemo9_Jsp 项目的结构如图 9-2 所示。

在 MySQL 数据库中建立数据库 springdemo9，在数据库内新建项目 8 中任务 8.9 案例中的 employee 数据表，内部添加 2 条数据。employee 表如图 9-3 所示，其中 emp_gender 为 0 表示男生，emp_gender 为 1 表示女生。下面就可以开始编写控制器类和 JSP 前端页面来实现增、删、改、查功能。

9.2.3 编写控制器类和 JSP 前端页面实现增、删、改、查

1. 查询所有员工

在 controller 目录下新建 EmployeeController 类，内部编写查询员工请求处理方法。

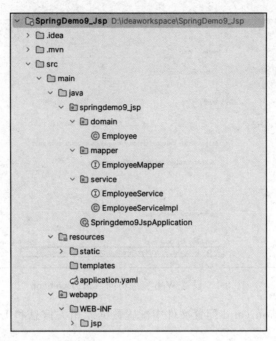

图 9-2　SpringDemo9_Jsp 项目结构

	emp_id	emp_name	emp_gender	emp_dept	emp_birth
1	1	员工1	0	部门1	2023-05-17
2	2	员工2	1	部门2	2023-05-16

图 9-3　employee 数据表

```
package springdemo9_jsp.controller;
import org.springframework.beans.factory.annotation.Autowired;
import org.springframework.stereotype.Controller;
import org.springframework.ui.Model;
import org.springframework.web.bind.annotation.*;
import springdemo9_jsp.domain.Employee;
import springdemo9_jsp.service.EmployeeService;
@Controller
public class EmployeeController {
    @Autowired
    private EmployeeService employeeService;
    @GetMapping("/emp") //查询所有员工
    public String select_emp(Model model) {
        //利用 model 传递所有员工数据,跳转到 WEB-INF/jsp/select_emp.jsp 页面
        model.addAttribute("emplist",employeeService.select_emp());
        return "select_emp";
    }
}
```

在 WEB-INF/jsp 目录下新建 select_emp.jsp 页面，内部通过<c:forEach>标签和 EL 表达式获取所有员工数据，并以表格展示。表格上方设置添加员工按钮，URL 为/toAddPage。表

格内每条记录后面设置修改员工和删除员工按钮,URL 分别为/toEditPage/${emp.emp_id}和/deleteEmp/${emp.emp_id}。

【select_emp.jsp】

```jsp
<%@ page contentType="text/html;charset=UTF-8" language="java"%>
<%@taglib prefix="c" uri="http://java.sun.com/jsp/jstl/core" %>
<html>
<head>
    <title>select emp</title>
</head>
<body>
<a href="/toAddPage">添加员工</a>
<table width="100%" border="1" cellspacing="1" cellpadding="0">
    <tr>
        <td>编号</td>
        <td>姓名</td>
        <td>性别</td>
        <td>部门</td>
        <td>出生日期</td>
        <td>操作</td>
    </tr>
    <c:forEach var="emp" items="${emplist}">
    <tr>
        <td>${emp.emp_id}</td>
        <td>${emp.emp_name}</td>
        <c:if test="${emp.emp_gender==0}">
            <td>男</td>
        </c:if>
        <c:if test="${emp.emp_gender==1}">
            <td>女</td>
        </c:if>
        <td>${emp.emp_dept}</td>
        <td>${emp.emp_birth}</td>
        <td><a href="/toEditPage/${emp.emp_id}">修改员工</a>
            <a href="/deleteEmp/${emp.emp_id}">删除员工</a>
        </td>
    </tr>
    </c:forEach>
</table>
</body>
</html>
```

运行项目启动类 Springdemo9JspApplication,在浏览器中输入地址 http://localhost:8080/emp,浏览器跳转到 select_emp.jsp 页面查询所有员工信息,如图 9-4 所示。

编号	姓名	性别	部门	出生日期	操作
添加员工					
1	员工1	男	部门1	2023-05-17	修改员工 删除员工
2	员工2	女	部门2	2023-05-16	修改员工 删除员工

图 9-4 查询所有员工信息

2. 添加员工

添加员工需单击表格上方的"添加员工"按钮,跳转到添加员工界面。添加成功则跳转到 select_emp.jsp 页面并展示添加的员工信息,添加失败则跳转回当前界面并提示添加失败。

在 WEB-INF/jsp 目录下新建 add_emp.jsp 页面,内部添加如下代码。必须保证每个字段都有 name 属性。

【add_emp.jsp】

```jsp
<%@ page contentType="text/html;charset=UTF-8" language="java" %>
<%@taglib prefix="c" uri="http://java.sun.com/jsp/jstl/core" %>
<html>
<head>
    <title>add emp</title>
</head>
<body>
<p>
<%-- 如添加用户失败,利用 msg 返回失败消息   --%>
<c:if test="${msg!=null}">
    ${msg}
</c:if>
</p>
<form action="/addEmp" method="post">
    <p>姓名:<input name="emp_name" type="text"></p>
    <p>性别:
        <select name="emp_gender">
        <option value="0">男</option>
        <option value="1">女</option>
        </select>
    </p>
    <p>部门:<input name="emp_dept" type="text"></p>
    <p>出生日期:<input name="emp_birth" type="text"></p>
    <input type="submit" value="新增">
</form>
</body>
</html>
```

在 EmployeeController 类中添加 URL 为/toAddPage 的请求处理方法,跳转新增员工界面 add_emp.jsp。添加 URL 为/addEmp 的请求处理方法新增员工,如员工信息添加成功,重定向到查询页面查询最新数据,如添加失败则跳转回 add_emp.jsp 并通过 msg 变量传递失败信息。

```java
@RequestMapping("/toAddPage")                //跳转新增员工页面
public String toAddPage(){
    return "add_emp";
}
@PostMapping("/addEmp")                      //新增员工
public String add_emp(Employee employee,Model model) {
```

```
    String res=employeeService.add_emp(employee);
    if(res.equals("add success")){
        return "redirect:/emp";
    }else{
        model.addAttribute("msg",res);
        return "add_emp";
    }
}
```

运行项目启动类 Springdemo9JspApplication，在浏览器中输入地址 http://localhost:8080/emp，浏览器跳转到 select_emp.jsp 页面查询所有员工信息。单击"新增员工"按钮，页面跳转到新增员工页面，在新增员工界面新增员工 3 的相关信息后，单击"新增"按钮，如图 9-5 所示。

员工信息新增成功，页面跳转回 select_emp.jsp，新增的员工信息也查询出来了，如图 9-6 所示。

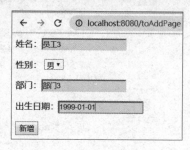

图 9-5 新增员工界面

图 9-6 新增员工成功

3. 修改员工

修改员工需单击表格中某条记录的"修改员工"按钮，跳转到修改员工界面。修改成功则跳转 select_emp.jsp 页面并展示修改后的员工信息，修改失败则不跳转界面并提示修改失败。

在 WEB-INF/jsp 目录下新建 edit_emp.jsp 页面，内部添加如下代码。修改员工需将员工原有的数据重新加载在页面上。此处用 EL 表达式取值，由于性别是下拉框，需用<c:if>标签进行条件判断以确定选择哪个 option。

```
<%@ page contentType="text/html;charset=UTF-8" language="java" %>
<%@taglib prefix="c" uri="http://java.sun.com/jsp/jstl/core" %>
<html>
<head>
    <title>edit emp</title>
</head>
<body>
<p>
<c:if test="${msg!=null}">
    ${msg}
</c:if>
${msg}
</p>
```

```jsp
<form action="/editEmp" method="post">
    <input name="emp_id" type="hidden" value=${emp.emp_id}>
    <p>姓名:<input name="emp_name" type="text" value="${emp.emp_name}"></p>
    <p>性别:
        <select name="emp_gender" >
            <%-- 根据 emp_gender 的值判断选择哪个 option --%>
            <c:if test="${emp.emp_gender==0}">
                <option value="0" selected="selected">男</option>
                <option value="1" >女</option>
            </c:if>
            <c:if test="${emp.emp_gender==1}">
                <option value="0">男</option>
                <option value="1" selected="selected">女</option>
            </c:if>
        </select>
    </p>
    <p>部门:<input name="emp_dept" type="text" value=${emp.emp_dept}></p>
    <p>出生日期:<input name="emp_birth" type="text" value=${emp.emp_birth}></p>
    <input type="submit" value="修改">
</form>
</body>
</html>
```

在 EmployeeController 类中添加 URL 为/toEditPage/{emp_id}的请求处理方法,跳转修改员工界面 edit_emp.jsp,方法内部利用 emp_id 查询该员工的信息并传递给 edit_emp.jsp 页面。添加 URL 为/editEmp 的请求处理方法修改员工,如员工信息修改成功则重定向到查询页面查询最新数据,如修改失败则重定向回 edit_emp.jsp 并通过 msg 变量传递失败信息。

```java
@RequestMapping("/toEditPage/{emp_id}")           //跳转修改员工页面
public String toEditPage(@PathVariable("emp_id") Integer emp_id,Model model){
    Employee employee=employeeService.select_empById(emp_id);
    model.addAttribute("emp",employee);
    return "edit_emp";
}
@PostMapping("/editEmp")                           //修改员工
public String update_emp(Employee employee,Model model) {
    String res=employeeService.update_emp(employee);
    if(res.equals("update success")){
        return "redirect:/emp";
    }else{
        model.addAttribute("msg",res);
        return "redirect:/toEditPage/"+employee.getEmp_id();
    }
}
```

运行项目启动类 Springdemo9JspApplication,在浏览器中输入地址 http://localhost:8080/emp,浏览器跳转到 select_emp.jsp 页面并查询所有员工信息。单击员工 3 的"修改员工"按钮,页面跳转到修改员工页面,将员工 3 的部门信息改为部门 4,再单击"修改"按钮,如图 9-7 所示。

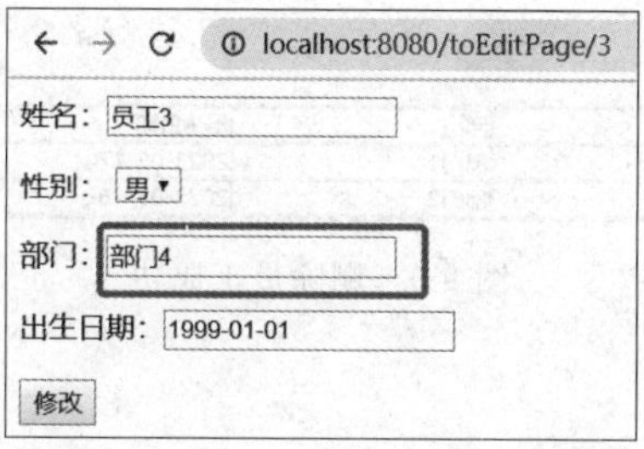

图 9-7 修改员工

员工信息修改成功,页面跳转回 select_emp.jsp,查询出员工 3 的部门被修改为部门 4,如图 9-8 所示。

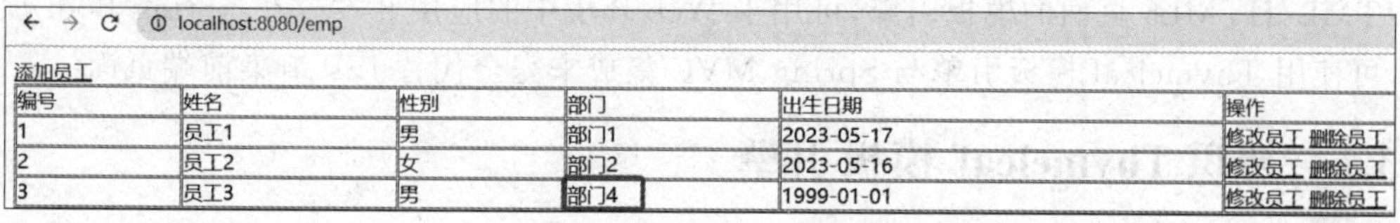

图 9-8 修改员工成功

4. 删除员工

删除员工需单击表格中某条记录的"删除员工"按钮,无论删除成功还是失败,都跳转 select_emp.jsp 页面查询员工信息;如果删除失败,则在 select_emp.jsp 页面提示删除失败。

可以在 select_emp.jsp 的＜body＞标签里面加入提示信息 msg。

```
<p>
<c:if test="${msg!=null}">
    ${msg}
</c:if>
${msg}
</p>
```

在 EmployeeController 类中添加 URL 为/deleteEmp/{emp_id}的请求处理方法,删除 emp_id 的员工,删除完毕,重定向到查询页面查询最新数据。如删除失败,则通过 msg 变量传递失败信息。

```
@GetMapping("/deleteEmp/{emp_id}")    //删除员工
public String delete_emp(@PathVariable("emp_id") Integer emp_id,Model model) {
    String res=employeeService.delete_emp(emp_id);
    model.addAttribute("msg",res);
    return "redirect:/emp";
}
```

运行项目启动类 Springdemo9JspApplication,在浏览器中输入地址 http://localhost:8080/emp,浏览器跳转到 select_emp.jsp 页面查询所有员工信息。单击员工 3 的"删除员工"按钮,删除完毕,页面跳转回 select_emp.jsp,可以看到员工 3 已经被成功删除了,如图 9-9 所示。

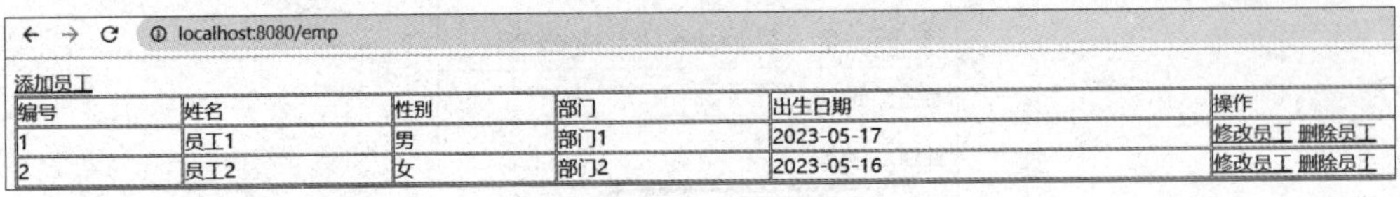

图 9-9 删除员工成功

任务 9.3 利用 Thymeleaf 模板引擎实现前端功能

Thymeleaf 是 Spring Boot 默认推荐的模板引擎。Thymeleaf 是一个基于 XML/XHTML/HTML5 页面的模板引擎,可用于 Web 环境中的应用开发。在 Spring Boot 项目中,可使用 Thymeleaf 模板引擎与 Spring MVC 集成来完全代替 JSP 渲染前端页面。

9.3.1 初识 Thymeleaf 模板引擎

Thymeleaf 是新一代 Java 模板引擎,Thymeleaf 基于 HTML 页面,以 HTML 标签为载体。Thymeleaf 既可以直接使用浏览器打开,查看页面的静态效果,也可以通过 Web 应用程序进行访问,查看动态页面效果。使用时,Thymeleaf 可与 Spring MVC 集成,快速地实现前后端数据交互。Thymeleaf 是 Spring Boot 默认推荐的模板引擎,Spring Boot 官方也推荐使用 Thymeleaf 来替代 JSP。

9.3.2 Spring Boot 引入 Thymeleaf 模板引擎

在 Spring Boot 中引入 Thymeleaf,只需在 pom.xml 文件中添加如下 Thymeleaf 依赖。

```xml
<dependency>
    <groupId>org.springframework.boot</groupId>
    <artifactId>spring-boot-starter-thymeleaf</artifactId>
</dependency>
```

Spring Boot 会自动对 Thymeleaf 进行配置。如配置默认的视图前缀为 classpath:/templates/,默认的视图后缀为.html。由此可以看出,Thymeleaf 模板引擎的 HTML 页面默认放置位置和 JSP 不同,为 resources 目录下的 templates 目录,该目录的 Spring Boot 已自动创建。

9.3.3 Thymeleaf 语法

使用 Thymeleaf,需先在 HTML 页面的<html>标签中声明名称空间,才可以在 HTML 页面中使用 Thymeleaf 语法。

```html
<html xmlns:th="http://www.thymeleaf.org">
```

Thymeleaf 通过在 HTML 标签属性中增加额外属性(原有 HTML 标签属性前增加 th 前缀)并结合 Thymeleaf 表达式取值实现"模板+数据"的页面渲染。Thymeleaf 常用的表达式如表 9-1 所示。

表 9-1　Thymeleaf 常用的表达式

表 达 式	语法	描　　述
变量表达式	${...}	获取请求域、session 域、对象等变量值
选择变量表达式	*{...}	变量表达式的对象取的是整个上下文中的变量,而选择表达式的对象取的是当前对象中的变量
消息表达式	#{...}	用于读取.properties 属性文件中的值,一般用于国际化
链接表达式	@{...}	用于生成资源链接,如引入静态资源及设置 URL 路由等
片段表达式	~{...}	类似于 jsp:include 的作用,用于引入公共页面片段

其中,变量表达式、链接表达式和片段表达式使用最多。变量表达式用于获取后端传值;链接表达式用于引入资源,设置 URL 路由;片段表达式用于组件化重构页面,减少重复代码。下面主要介绍这三类表达式的语法。

1. 变量表达式

变量表达式用于获取后端传值,根据传值的类型不同,前端可以使用不同的变量表达式接收。

(1) 接收简单类型数据。假设后端控制器使用 Model 对象传值。

```
model.addAttribute("username","user2");
```

前端 HTML 页面可使用 ${变量名} 接收数据。<p>、<span> 等文本类标签使用 th:text 属性接收数据,input 输入框使用 th:value 属性接收数据。

```
<p th:text="${username}"></p>或<input th:value="${username}">
```

(2) 接收 Java 对象。假设后端控制器使用 Model 对象传递 User 对象,User 对象内部有 username 和 password 属性。

```
User user=new User("user1","123456");
model.addAttribute("user",user);
```

前端 HTML 页面可使用 ${参数名.属性名} 变量表达式接收。

```
<p th:text="${user.username}"></p>或<input th:value="${user.username}">
<p th:text="${user.password}"></p>或<input th:value="${user.password}">
```

(3) 接收 Java 对象集合。假设后端控制器使用 Model 对象传递 User 对象集合,User 对象内部有 username 和 password 属性。

```
User user1=new User("user1","123456");
User user2=new User("user2","56789");
```

```
List<User> userList=new ArrayList<User>();
userList.add(user1);
userList.add(user2);
model.addAttribute("userlist",userList);
```

前端 HTML 页面可使用 th:each="user:${userlist}"变量表达式接收数据,其中 userlist 对应后端传递参数名。user 对应集合对象里面每个元素。此处将 username 和 password 以拼接字符串显示。

```
<p th:each="user:${userlist}" th:text="${user.username}+','+${user.password}"></p>
```

或

```
<input th:each="user:${userlist}" th:value="${user.username}+','+${user.password}">
```

(4) 接收 Session 对象数据。假设后端控制器使用 Session 对象传递 User 对象,User 对象内部有 username 和 password 属性。

```
session.setAttribute("user",user);
```

前端 HTML 页面可使用 ${session.参数名.属性名}接收数据。此处将 username 和 password 以拼接字符串显示。

```
<p th:text="${session.user.username}+','+${session.user.password}"></p>
```

或

```
<input th:value="${session.user.username}+','+${session.user.password}">
```

2. 链接表达式

链接表达式用于引入资源,设置 URL 路由。

(1) 引入静态资源。如在 Spring Boot 项目中,静态资源放置在 static 目录下的 css、js 和 img 等目录,如果页面需要引入静态 JS 资源和图片资源,可以采用 th:src 属性,内部利用@{}将资源路径包裹起来。

```
<script th:src="@{/js/1.js}"></script>        //引入 static/js/1.js 文件
<img th:src="@{/img/1.jpg}" alt="">           //引入 static/img/1.jpg 文件
```

如需引入静态 CSS 资源,可以采用 th:href 链接表达式。

```
<link th:href="@{/css/1.css}" rel="stylesheet">  //引入 static/css/1.css 文件
```

(2) 超链接设置 URL 路由。假设在 HTML 页面中设置一些超链接用于页面跳转,可以采用 th:href 属性,内部利用@{}将 URL 路由包裹起来。URL 路由使用字符串拼接,同时还可传递参数。

```
<a th:href="@{'/editUser/'+${userId}}">修改用户</a>
```

（3）表单提交设置 URL 路由。表单提交数据也是页面提交数据的常用方式，表单内有 action 属性用于设置提交的 URL 路由。可以使用 th:action 属性设置 URL，内部利用 @{} 将 URL 路径包裹起来。

```
<form th:action="@{/subscribe}">
```

3. 片段表达式

片段表达式一般用于页面的模块化引入，将各个页面中的公共部分抽取出来，形成一个 HTML 页面，然后引入其他页面中，这样能够减少页面重复代码，增强可维护性。

假设有页脚 foot.html，需要在首页 index.html 下部引入，可将 foot.html 页面中需要引入的部分设置 th:fragment 属性。

```
<body>
    <div th:fragment="sub">页面页脚</div>
</body>
```

在首页 index.html 中采用 th：insert 属性引入，th：insert 属性值用 ~{}包裹。

```
<body>
    <p>这是首页 index</p>
    <div th:insert="~{foot :: sub}"></div>
</body>
```

除了利用表达式直接取值之外，Thymeleaf 也支持在表达式内部进行条件判断取值。例如性别为 0 时输出男，性别为 1 时输出女，可以采用如下写法利用三目运算符条件判断。

```
<p th:text="${user.age eq 0}?'男':'女'"></p>
```

Thymeleaf 也支持 if 条件判断取值，但不支持 if else 判断。如在页面显示相应信息时，经常会做如下判断，如果 msg 不为空，则显示内容，否则不显示。可以采用如下写法。

```
<p th:if="${null != msg}">${msg}</span>
```

Thymeleaf 还提供了众多的工具类表达式用于各种类型的数据处理。例如，#strings 用于处理字符串数据，#dates 用于处理日期数据，#lists 用于处理 list 集合数据等。开发中常用的一些工具类表达式如下。

```
${#strings.isEmpty(str)}                              //字符串是否为空
${#strings.length(str)}                               //获取字符串长度
${#strings.trim(str)}                                 //去除空格
${#strings.equals(first, second)}                     //判断两个字符串是否相等
${#dates.format(date, 'yyyy-MM-dd hh:mm:ss')}         //时间日期格式化
```

```
${#dates.createNow()}                          //获取系统当前时间
${#dates.createToday()}                        //获取系统当前日期
${#lists.isEmpty(list)}                        //list 集合是否为空
${#lists.contains(list, element)}              //list 集合是否包含某个元素
${#aggregates.sum(collection)}                 //聚合函数,求集合元素和
${#aggregates.avg(collection)}                 //聚合函数,求集合元素平均值
```

注意:如果在 Thymeleaf 页面向服务端请求传递数据,必须保证页面所传递的数据有 name 属性,否则控制器无法通过 name 获取数据。如页面内容如下。

```
<p>用户名:<input name="username" type="text"></p>
<p>密码:<input name="password" type="text"></p>
```

控制器才能够通过(String username,String password)获取请求数据。

9.3.4 编写 Thymeleaf 前端页面实现增、删、改、查

这里重新创建一个新的 Spring Boot 项目 SpringDemo9_Thymeleaf,还是以员工信息管理为例,演示编写 Thymeleaf 前端页面实现增、删、改、查。在项目中后端代码同本项目任务 9.2 的相同,即复用该任务的持久层、业务层和控制层代码,这里只实现 Thymeleaf 前端页面。

在 SpringDemo9_Thymeleaf 项目的 pom.xml 文件中引入如下依赖。

```xml
<!--引入 thymeleaf-->
<dependency>
    <groupId>org.springframework.boot</groupId>
    <artifactId>spring-boot-starter-thymeleaf</artifactId>
</dependency>
<dependency>
    <groupId>org.springframework.boot</groupId>
    <artifactId>spring-boot-starter-web</artifactId>
</dependency>
<dependency>
    <groupId>org.projectlombok</groupId>
    <artifactId>lombok</artifactId>
    <optional>true</optional>
</dependency>
<!-- 引入 Mybatis、MySQL 驱动和 Druid-->
<dependency>
    <groupId>org.mybatis.spring.boot</groupId>
    <artifactId>mybatis-spring-boot-starter</artifactId>
    <version>3.0.1</version>
</dependency>
<dependency>
    <groupId>com.mysql</groupId>
    <artifactId>mysql-connector-j</artifactId>
</dependency>
<dependency>
```

```xml
        <groupId>com.alibaba</groupId>
        <artifactId>druid-spring-boot-starter</artifactId>
        <version>1.2.16</version>
    </dependency>
    <dependency>
        <groupId>org.springframework.boot</groupId>
        <artifactId>spring-boot-starter-test</artifactId>
        <scope>test</scope>
    </dependency>
```

在项目的 application.yaml 配置文件中配置数据库相关信息,并关闭 thymeleaf 页面缓存,以便调试页面。thymeleaf 页面缓存开发环境不建议开启,生产环境下可打开,可增加页面渲染速度。

```yaml
spring:
  #数据库配置
  datasource:
    url: jdbc:mysql://localhost:3306/springdemo9?&useUnicode=true&characterEncoding=utf8&serverTimezone=UTC
    username: root
    password: 123456
    driver-class-name: com.mysql.cj.jdbc.Driver
    type: com.alibaba.druid.pool.DruidDataSource
  #关闭 thymeleaf 缓存
  thymeleaf:
    cache: false
```

下面开始编写 thymeleaf 页面。

1. 查询员工和删除员工

编写查询员工页面 select_emp.html。查询页面包含删除员工功能,并将删除结果显示在页面上部。

【select_emp.html】

```html
<!DOCTYPE html>
<html lang="en" xmlns:th="http://www.thymeleaf.org">
<head>
    <meta charset="UTF-8">
    <title>select emp</title>
</head>
<body>
<!--获取删除员工结果,如果有值则显示-->
<p th:if="${ msg != null}" th:text="${ msg}"></p>
<a th:href="@{/toAddPage}">添加员工</a>
<table width="100%" border="1" cellspacing="1" cellpadding="0">
    <tr>
        <td>编号</td>
        <td>姓名</td>
        <td>性别</td>
```

```html
            <td>部门</td>
            <td>出生日期</td>
            <td>操作</td>
        </tr>
        <!-- 遍历emplist获取员工信息-->
        <tr th:each="emp: ${emplist}">
            <td th:text="${emp.emp_id}"></td>
            <td th:text="${emp.emp_name}"></td>
            <!-- 根据性别0或1显示男、女-->
            <td th:text="${emp.emp_gender eq '0'}? '男':'女'"></td>
            <td th:text="${emp.emp_dept}"></td>
            <td th:text="${emp.emp_birth}"></td>
            <td>
                <a th:href="@{'/toEditPage/'+${emp.emp_id}}">修改员工</a>
                <a th:href="@{'/deleteEmp/'+${emp.emp_id}}">删除员工</a>
            </td>
        </tr>
    </table>
</body>
</html>
```

2. 新增员工

编写新增员工页面 add_emp.html。必须保证每个字段都有 name 属性。

【add_emp.html】

```html
<!DOCTYPE html>
<html lang="en" xmlns:th="http://www.thymeleaf.org">
<head>
    <meta charset="UTF-8">
    <title>add emp</title>
</head>
<body>
<!--获取添加员工结果,如果有值则显示-->
<p th:if="${ msg != null}" th:text="${ msg}"></p>
<form th:action="@{/addEmp}" method="post">
    <p>姓名:<input name="emp_name" type="text"></p>
    <p>性别:
        <select name="emp_gender">
            <option value="0">男</option>
            <option value="1">女</option>
        </select>
    </p>
    <p>部门:<input name="emp_dept" type="text"></p>
    <p>出生日期:<input name="emp_birth" type="text"></p>
    <input type="submit" value="新增">
</form>
</body>
</html>
```

3. 修改员工

编写修改员工页面 edit_emp.html。修改某个 emp_id 的员工信息，必须保证每个字段都有 name 属性，且通过表达式接收该员工的当前信息。

【edit_emp.html】

```html
<!DOCTYPE html>
<html lang="en" xmlns:th="http://www.thymeleaf.org">
<head>
    <meta charset="UTF-8">
    <title>edit emp</title>
</head>
<body>
<!--获取修改员工结果,如果有值则显示-->
<p th:if="${ msg != null}" th:text="${ msg}"></p>
<form th:action="@{/editEmp}" method="post">
    <input name="emp_id" type="hidden" th:value="${emp.emp_id}">
    <p>姓名:<input name="emp_name" type="text"
            th:value="${emp.emp_name}"></p>
    <p>性别:
        <select name="emp_gender">
            <!-- 根据 emp_gender 的值判断选择哪个 option-->
            <option value="0" th:selected="${emp.emp_gender eq '0'}">男</option>
            <option value="1" th:selected="${emp.emp_gender eq '1'}">女</option>
        </select>
    </p>
    <p>部门:<input name="emp_dept" type="text" th:value="${emp.emp_dept}"></p>
    <p>出生日期:<input name="emp_birth"
            type="text" th:value="${emp.emp_birth}"></p>
    <input type="submit" value="修改">
</form>
</body>
</html>
```

select_emp.html、add_emp.html 和 edit_emp.html 页面操作流程和本项目任务 9.2 中的 JSP 页面一样，这里不再展示，读者可以自行测试。

任务 9.4 利用 Vue 实现前端功能

目前 Web 应用开发的主流是前后端分离开发模式，在前后端分离开发模式中，Vue 因其使用简单且易于上手而成为当下使用最广泛的前端框架之一。本节将基于 Vue3 介绍如何编写一个独立的 Vue 前端项目和 Spring Boot 后端进行数据交互，实现 Web 应用的前后端分离开发。

9.4.1 初识 Vue

传统的 Web 前端开发大多是基于原生 JS/jQuery 模板引擎的方式来构建用户界面，且前后端代码都存在于同一个项目中。这种开发模式一方面使得项目前后端高度耦合，不利于后续的代码维护和功能扩展；另一方面在前后端数据交互时需要频繁地操作页面 DOM 元素来加载最新的数据，使前端开发人员把大量精力和时间花在 DOM 元素的反复操作上，而不能专注于处理业务逻辑，影响开发效率和页面性能。

Vue 介绍及 MVVM 原理

Vue 是一套用于构建用户界面的前端框架。Vue 框架能够很好地解决上述问题，从而提高前端开发效率和页面性能。Vue 的优势主要体现在如下几个方面。

（1）Vue 能够构建一个独立的前端项目，项目内部存放所有前端资源，实现 Web 应用的前后端分离开发，减少了前后端代码的耦合度，方便后续代码维护和功能扩展。

（2）与 React 和 Angular 等前端框架相比，Vue 足够轻量，学习成本也较低。

（3）Vue 采用基于面向组件的思维能够快速构建单页面应用程序。整个 Vue 项目只有一个首页，一般为 index.html，可以通过在首页挂载不同的 Vue 组件实现页面 DOM 的切换。

（4）Vue 组件基于（model-view-viewmodel，MVVM）思想实现页面 DOM 和数据的双向绑定。Vue 中的数据是响应式数据，前端开发人员可以通过 Vue 提供的相关指令实现响应式数据和 DOM 的双向绑定。即页面 DOM 能够自动监听数据的变化并渲染页面，同时如有用户操作改变页面 DOM 数据值，Vue 也能及时获取最新数据，这样能够使前端开发人员脱离复杂的 DOM 操作，专注处理业务逻辑。

（5）除了双向数据绑定的核心功能之外，Vue 还提供一整套组件和项目扩展工具，包括路由跳转组件（vue-router）、状态管理组件（vuex）、模板组件（component）、项目构建脚手架工具（vue-cli）和浏览器调试工具（vue-devtools）等，以使前端开发人员能够更方便地构建和开发 Vue 项目。

MVVM 思想是 Vue 实现页面 DOM 和数据双向绑定的核心思想。MVVM 思想将一个 HTML 页面拆分成了 Model、View 和 ViewModel 三部分，如图 9-10 所示。

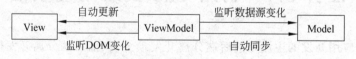

图 9-10　MVVM 工作原理

其中，View 表示当前页面所渲染的 DOM 结构，Model 表示当前页面渲染时所依赖的数据源。ViewModel 表示 Vue 的实例，ViewModel 把当前页面的 Model 和 View 绑定在一起，实现了页面 DOM 和数据双向绑定。当 Model 发生变化时，会被 ViewModel 监听到，

ViewModel 会根据最新的数据源自动更新页面的结构。当 View 的值发生变化时,也会被 ViewModel 监听到,ViewModel 会把 View 最新的值自动同步到 Model 中。

目前 Vue 的主流开发版本是 3.0 版本。本书后续将基于 3.0 版本介绍 Vue 的安装、配置和使用。

9.4.2 搭建 Vue3 开发环境

搭建 Vue 开发环境的主要步骤如下。

1. 安装 node.js

与传统的 Web 应用一样,Vue 项目也有自己的运行环境。传统的 Web 应用运行需借助 Web 服务器实现,例如 Tomcat、WebLogic、JBoss 等。Vue 运行需借助 node.js 实现。读者可自行从 node.js 官网下载最新的 LTS 版本的 node.js 安装。Vue3 要求 node.js 版本为 16.0 版本或以上,本书使用的 node.js 版本为 18.16 版本。node.js 的安装步骤这里不详细介绍,安装成功后可以在命令行终端输入 node -v 命令,查询 node.js 版本。

2. 安装 VsCode 并添加 Volar 插件

Vue 作为一个前端项目也可以使用 Idea 来开发,但是还有很多更轻量级的开发工具可以选择,如 VsCode、HBuilder、WebStrom、Sublime Text 等。其中 VsCode 使用者最多,本书也选择 VsCode 来开发 Vue 项目,因此需先安装 VsCode 软件。VsCode 软件可自行从 VsCode 官网下载最新的 Stable 版本安装,本书使用的 VsCode 版本为 1.78 版本。VsCode 安装方法这里不详细介绍。在 VsCode 安装完毕,还需在 VsCode 中安装 Volar 插件。Vue 团队官方推荐使用 Volar 插件进行 Vue3 代码的开发,Volar 插件支持 Vue3 语言高亮显示和语法提示,同时提供对 TypeScript 的完美支持。Volar 插件的安装非常简单,如图 9-11 所示,直接在 VsCode 的插件市场搜索 volar,然后单击 volar 右下角 Install 按钮安装就可以了。

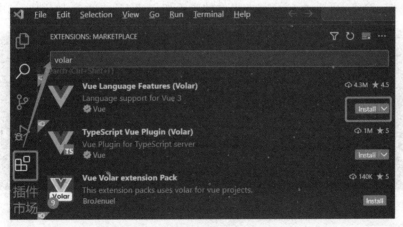

图 9-11 安装 Volar 插件

9.4.3 创建 Vue3 项目

Vue 官方推荐使用 npm init vue@latest 命令创建 Vue3 项目,该命令基于 Vite 创建

Vue3 项目,项目运行时使用 Vite 打包运行。Vite 是一个类似于 WebPack 的前端项目打包工具,与 WebPack 相比,基于 Vite 构建的项目结构更加简洁,打包运行速度更快。本小节将基于 Vite 创建一个 Vue3 项目,具体创建步骤如下。

(1) 在本地磁盘的 D 盘中新建一个存放 Vue 项目的文件夹,文件夹命名为 vue3demo。打开 VsCode,将 vue3demo 文件夹拖入 VsCode 中,VsCode 弹出一个界面,勾选复选框,并单击"Yes,I trust the authors"按钮设置信任,如图 9-12 所示。

图 9-12 设置信任

(2) 单击 VsCode 上方菜单栏中的 Terminal 按钮,下方弹出终端窗口,窗口自动切换到 vuedemo3 目录下,在终端窗口中输入 npm init vue@latest 命令,按 Enter 键,则会安装并执行 create-vue 命令,再通过 Vue 官方提供的脚手架创建 Vue3 项目,如图 9-13 所示。

图 9-13 创建 Vue3 项目

(3) 在终端窗口中输入项目名称 Project name 为 vue001,依次设置所有 Add 选项都为 No,不添加任何可选功能,后续如需要添加,可以通过命令按需添加。按 Enter 键,终端窗口显示 vue001 done.Now run。此时 vue001 项目目录下文件已生成完毕,可以按照提示信息依次执行后续命令来添加 Vue3 相关依赖,如图 9-14 所示。

(4) 在终端窗口输入 cd vue001 命令,进入 Vue3 项目的根目录中。输入 npm install 命令,安装 Vue3 相关依赖,如图 9-15 所示。

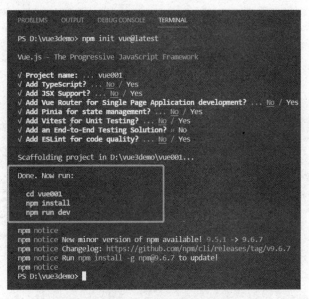

图 9-14 Vue 项目创建完成

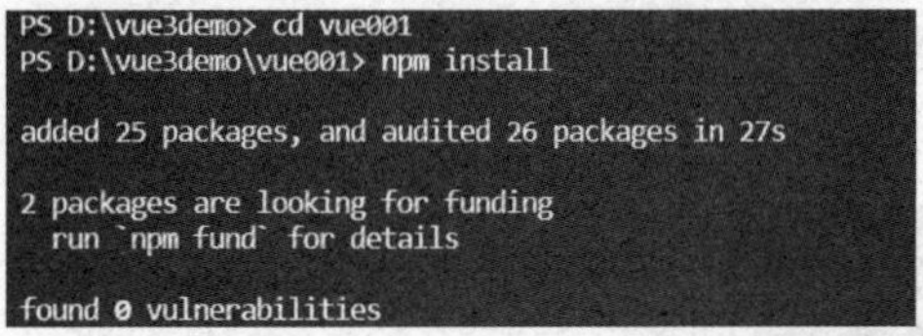

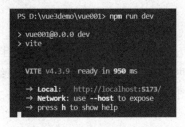

图 9-15 安装 Vue3 相关依赖

(5) 等待一小会儿,待所有依赖安装完毕,单击 package.json 文件,在 package.json 文件内部可以看到如下信息,Vue 的版本为 3.3.2。

```
"dependencies": {
    "vue": "^3.3.2"
},
```

再输入 npm run dev 命令运行 Vue3 项目。运行成功后,终端显示浏览器访问地址为 http://localhost:5173/,如图 9-16 所示。

图 9-16 成功运行 Vue3 项目

(6) 在浏览器地址栏中输入 http://localhost:5173/,显示 Vue3 项目首页,如图 9-17

所示。

图 9-17　Vue3 项目首页

9.4.4　Vue3 项目目录结构及访问机制

下面简单介绍基于 Vite 构建的 Vue3 项目目录结构。整个项目目录结构如图 9-18 所示。

图 9-18　项目目录

Vue3 项目目录结构及访问机制

其中，各目录作用如下。

node_modules：放置通过 npm install 下载并安装的项目依赖包。

public：存放静态资源公共资源，项目打包运行不会被压缩合并。

——favicon.ico：项目网站图标。

src：项目开发主要目录。

——assets：放置项目静态文件、图片等资源。

——components：Vue 组件默认放置的位置，也可用于存放个人开发的 Vue 组件。

—App.vue：项目的根组件。
—main.js：项目入口.JS 文件，默认挂载 App.vue 跟组件。
.gitignore：配置哪些文件不归 git 管理。
index.html：项目首页入口.html 文件，内部引入 main.js。
package-lock.json：锁定通过 npm install 下载并安装的项目依赖包版本号，使每次下载依赖版本号保持一致。
package.json：项目配置和依赖包管理。
README.md：说明文档，主要看项目运行的命令。
vue.config.js：项目配置信息，如配置跨域 proxy 代理等。

Vue3 项目的访问机制为：当浏览器访问 Vue3 项目时，会首先访问 index.html 页面。而 index.html 页面内部又引入 main.js，因此页面会加载 main.js 内部代码。

main.js 内部代码如下：

```
import './assets/main.css'
import { createApp } from 'vue'
import App from './App.vue'
createApp(App).mount('#app')
```

该代码使用了 Vue3 中的 createApp 函数传入项目根组件 App.vue 作为参数，创建一个 Vue3 应用实例，并将该实例挂载到 index.html 页面中 id 为 app 的 DOM 元素上，同时将 src/assets/ 目录下 main.css 中的样式全局作用于 id 为 app 的 DOM 上，至此才能在 Vue3 项目首页看到 App.vue 内部的内容。App.vue 作为父组件存在，内部又引入了 HelloWorld 和 TheWelcome 两个子组件，并向子组件 HelloWorld 传递了 msg 参数。

```
<template>
  <header>
    < img alt="Vue logo" class="logo" src="./assets/logo.svg" width="125" height="125" />
    <div class="wrapper">
      <HelloWorld msg="You did it!" />
    </div>
  </header>
  <main>
    <TheWelcome />
  </main>
</template>
```

在 Vue3 项目中，页面内容通常以.vue 组件形式存在，以便实现模块化复用。一个.vue 组件的基本结构如下：

```
<template></template>            //内部添加 HTML 元素,也可以引入其他 Vue 组件
<script></script>                //内部可添加页面 JS 代码
<style></style>                  //部可添加页面 CSS 样式代码
```

组件内部包含<template>标签、<script>标签和<style>标签。其中，<template>

标签是组件的视图模板层，内部可添加 HTML 元素，也可以引入其他 Vue 组件。<script>标签是组件的数据和方法层，内部可添加页面 JS 代码，包括 import 引入资源、定义变量和方法等。<style>标签是组件的样式层内部可添加页面 CSS 样式代码。

9.4.5 Vue3 组件入口函数——setup 函数

当页面加载一个 Vue 组件时，通常会通过一个入口函数来访问组件内容。Vue3 中提供 setup 函数作为所有组件的统一入口函数。setup 函数是一个组合式 API 入口函数，定义在组件的<script>标签中。setup 函数在创建当前组件实例时会自动执行，是组件内部所有函数中最早执行的。定义了 sctup 函数后，组件中所有变量和方法都必须写在 setup 函数中且必须通过 return 返回出去才可以在视图层中访问。

这里在 vue001 项目 src/components 目录下新建 My.vue 文件，内部输入如下代码。

```
<template>
  <div id="app">
    <p>{{ hello }}</p>                    //插值表达式显示 hello 变量的值
  </div>
</template>
<script>
export default {
  setup() {
    let hello ="Hello Vue3";              //在 setup 函数中定义一个变量
    return {
      hello,                              //hello 必须通过 return 返回出去
    };
  },
};
</script>
```

上述代码定义了一个变量 Hello 并赋值为 Hello Vue3，该变量通过插值表达式{{ hello }}，将值显示在页面上。插值表达式的使用语法为{{变量名}}。

修改项目 mian.js 文件代码，将 id 为 app 的 DOM 上挂载的 Vue 组件修改为 My.vue。

```
import './assets/main.css'
import { createApp } from 'vue'
import App from './App.vue'
import my from './components/My.vue'
createApp(my).mount('#app')
```

在 VsCode 终端中切换到 vue001 目录下，输入 npm run dev 命令启动项目，在浏览器中输入访问地址 http://localhost:5173/，页面显示 Hello Vue3，实现了数据渲染视图。

下面在 My.vue 中再新增一个按钮并添加 Click 事件，用于改变 hello 变量的值。

```
<template>
    <p>{{ hello }}</p>
    <button @click="getHelloValue">获取值</button>
</template>
```

```
<script>
export default {
    setup() {
        let hello = "Hello Vue3";
        function getHelloValue(){
            console.log(hello);
        }
        return {
            hello,
            getHelloValue, //getHelloValue 必须通过 return 返回出去
        };
    },
};
</script>
```

上述代码定义了一个按钮"获取值"并添加了 Click 事件,执行 getHelloValue 方法用于获取 hello 变量的值。getHelloValue 方法内部控制台打印输出 hello 的值。

保存代码后刷新页面,页面显示 hello 初始值为 Hello Vue3,然后单击"获取值"按钮,控制台打印出 hello 值为 Hello Vue3,方法正常执行完毕。

在编程时,setup 函数可以简写为＜script setup＞＜/script＞,这样＜script＞标签内部的所有代码会被编译成 setup 函数的内容,在每次组件实例被创建的时候执行,同时任何在＜script setup＞中声明的变量,函数和 import 引入的内容都能在视图模板中直接使用,而不需要 return 返回出去。

setup 函数内部还带有 props 和 context 两个参数。其中,props 用于接收外界传递的参数,接收数据格式为 props.变量名,一般用于父子组件的传值。context 是一个上下文对象,内部包含了 attrs、slots 等属性,也可用于父子组件的传值,外界传递的参数如果没有通过 props 接收,默认都会全部传递给 context 对象接收,context 对象接收数据格式为 context.attrs.变量名。下面简单演示父子组件传值。

在 components 目录下定义子组件 Sub.vue,内部添加如下代码来接收父组件传递的 msg 变量。

【Sub.vue】

```
<template>
<p>mysub</p>
</template>
<script>
export default {
    //采用 setup 函数的 props 属性接收的是响应式数据,必须使用 props 配置接收的数
    //据类型,才能解构获取数据,否则接收不到数据
    props:{
        msg:{
            type:String
        }
    },
    setup(props,context){
```

```
        //控制台打印通过 props 和 context 接收的数据
        console.log('mysub props:'+props.msg);
        console.log('mysub context:'+context.attrs.msg);
    }
}
</script>
```

在上述代码中,由于 props 接收的是响应式数据,需先指定 msg 的类型才能解构取值,这里指定 msg 类型为 String。而 context 对象接收的不是响应式数据,可以直接通过解构取值。

在父组件 My.vue 中添加如下黑体代码。

```
<template>
    <p>{{ hello }}</p>
    <button @click="getHelloValue">获取值</button>
    //页面使用子组件,传递 hello 变量的值给子组件,msg 前面一定要带冒号
    <Sub :msg="hello"></Sub>
</template>
<script>
//引入 Sub.vue 子组件资源
import Sub from './Sub.vue';
export default {
    components: { Sub },
    setup() {
        let hello = "Hello Vue3";
        function getHelloValue() {
            console.log(hello);
        }
        return {
            hello,
            getHelloValue,
        };
    },
};
</script>
```

上述代码中父子组件传值使用,表示父组件向子组件<Sub>标签里 msg 变量传递的值为 hello 变量的值 Hello Vue3。如果去掉冒号,写成,则父组件向子组件<Sub>标签里 msg 变量传递的值为 hello 字符串。

启动 Vue 项目,在谷歌浏览器中输入访问地址 http://localhost:5173/,浏览器控制台打印输出如下语句,props 中接收的数据为 Hello Vue3,props 接收后数据将不传递到 context 对象中。context 对象接收的 msg 为 undefined。

```
mysub props:Hello Vue3
mysub context:undefined
```

如果要使用 context.attrs.msg 接收数据，必须删除 Sub.vue 中的 props 代码，不使用 props 参数接收数据，这样父组件发送的 msg 数据才会传递给 context 对象接收。

```
props:{
    msg:{
        type:String
    }
},
```

9.4.6 创建和渲染响应式数据

在 Vue 框架中，页面 DOM 和数据实现了双向绑定。双向的含义如下。
- 数据→页面方向：Vue 中数据值改变能够被页面 DOM 自动监听并更新渲染页面。
- 页面→数据方向：如有用户操作改变页面 DOM 数据值，Vue 中的数据值也能及时更新。

在双向数据绑定中，响应式数据是实现双向数据绑定的基础。上述案例 My.vue 中定义的变量 hello 仅仅是一个字符串变量，而不是一个响应式数据。虽然 hello 的值能够显示在页面上，但是如果 hello 值变化，页面数据是不会及时更新的。这里在 My.vue 中新增一个"改变值"按钮并添加 click 事件，单击该按钮，将变量 hello 的值改为 Hello Vue4。

```
<template>
    <p>{{ hello }}</p>
    <button @click="getHelloValue">获取值</button>
    <Sub :msg="hello"></Sub>
    <button @click="changeValue">改变值</button>
</template>
<script>
import Sub from './Sub.vue';
export default {
    components: { Sub },
    setup() {
        let hello = "Hello Vue3";
        function getHelloValue() {
            console.log(hello);
        }
        function changeValue() {
            hello = "Hello Vue4";
            console.log(hello);
        }
        return {
            hello,
            getHelloValue,
            changeValue,
        };
```

```
        },
    };
</script>
```

启动项目,在浏览器访问地址 http://localhost:5173/,页面显示 hello 初始值为 Hello Vue3,然后单击"改变值"按钮,hello 的值被改变了,控制台打印出 hello 值为 Hello Vue4,但是页面数据并没有改变,还是显示 Hello Vue3。因此,要想使页面同步更新 hello 的值,hello 必须为响应式数据。

下面将介绍如何创建不同类型响应式数据并在页面渲染。

1. 创建响应式数据

Vue3 中可利用 ref 函数和 reactive 函数两种方式创建响应式数据。下面将详细介绍二者的使用。

(1) ref 函数。ref 函数一般用于创建简单数据类型的响应式数据,也可以用于创建对象类型的响应式数据。ref 简单数据类型响应式依靠 Object.defineProperty() 的 get 与 set 方法实现,对象类型的响应式底层将借助 reactive 函数实现。ref 函数使用示例如下。

```
<script>
import { ref } from 'vue';              //从 vue 对象中引入 ref 函数
export default {
    setup() {
        let hello = ref("a");            //创建一个 ref 响应式字符串数据,数据值为字符串 a
        //创建一个 ref 响应式对象数据,数据值为一个对象
        let obj=ref({
            a:1,
            b:2
        })
    },
};
</script>
```

ref 函数接受一个内部值并返回一个响应式且可变的 ref 对象,这个对象只包含一个 .value 属性。想要访问或改变 ref 对象的内部值,可使用 .value 属性。如对上述代码中 hello 和 obj 响应式数据取值和重新赋值,可使用如下代码。

```
hello.value                              //获取 hello 变量的值
hello.value="b";                         //对 hello 变量重新赋值
obj.value.a                              //获取 obj 对象 a 属性的值
obj.value.a=3;                           //对 obj 对象 a 属性重新赋值
```

下面对 My.vue 中的代码做修改,重新定义 hello 和 obj 两个 ref 响应式数据,单击"改变值"按钮,能够改变 hello 和 obj 的值,并在页面 DOM 中同步更新。

```
<template>
    <p>{{ hello }}</p>
    <p>{{ obj }}</p>
```

```
            <button @click="getHelloValue">获取值</button>
            <Sub msg="hello"></Sub>
            <button @click="changeValue">改变值</button>
    </template>
    <script>
    import { ref } from 'vue';
    import Sub from './Sub.vue';
    export default {
        components: { Sub },
        setup() {
            let hello = ref("Hello Vue3");        //ref 创建字符串类型响应式数据
            let obj=ref({a:1,b:2});               //ref 创建对象类型响应式数据
            function getHelloValue() {
                console.log(hello);
            }
            function changeValue() {
                hello.value = "Hello Vue4";       //改变响应式数据 hello 的值为 Hello Vue4
                obj.value.a=3;                    //改变响应式数据 obj 内部属性 a 的值为 3
                obj.value.b=4;                    //改变响应式数据 obj 内部属性 b 的值为 4
            }
            return {
                hello,
                obj,
                getHelloValue,
                changeValue,
            };
        },
    };
    </script>
```

启动项目，在浏览器输入访问地址 http://localhost:5173/，页面显示 Hello Vue3{ "a": 1, "b": 2}，然后单击"改变值"按钮，页面数据同步更新变为 Hello Vue4{ "a": 3, "b": 4}。

（2）reactive 函数。reactive 函数用于创建对象类型或数组类型的响应式数据。上述 ref 函数创建对象类型的响应式数据底层实现也借助了 reactive 函数。reactive 函数使用 Proxy 代理机制基于源对象创建了一个代理对象，代理对象通过 Reflect 反射机制对源对象内部数据进行操作。与 ref 函数相比，使用 reactive 实现的响应式数据在数据取值和赋值过程中省略了.value 属性，更加贴近编程习惯。reactive 函数使用示例如下。

```
let obj1=reactive({a:7,b:8});    //创建一个响应式对象数据 obj1,内部有 a 和 b 两个属性
let obj1=reactive([1,2,3]);      //创建一个响应式数组数据 obj1,内部 1、2、3 三个整型元素
```

下面对 My.vue 中的代码做修改，在原有基础上新增 reactive 响应式数据 obj1 和 obj2。其中，obj1 为对象类型响应式数据，obj2 为数组类型响应式数据。单击"改变值"按钮能够同时改变 hello、obj、obj1 和 obj2 的值，并在 DOM 页面同步更新。

```
<template>
    <p>{{ hello }}</p>
```

```
        <p>{{ obj }}</p>
        <p>{{ obj1 }}</p>
        <p>{{ obj2 }}</p>
        <button @click="getHelloValue">获取值</button>
        <Sub :msg="hello"></Sub>
        <button @click="changeValue">改变值</button>
</template>
<script>
import { reactive, ref } from 'vue';
import Sub from './Sub.vue';
export default {
    components: { Sub },
    setup() {
        let hello = ref("Hello Vue3");
        let obj=ref({a:1,b:2});
        let obj1=reactive({a:7,b:8});       //reactive 创建对象类型响应式数据
        let obj2=reactive([1,2,3]);         //reactive 创建数组类型响应式数据
        function getHelloValue() {
            console.log(hello);
        }
        function changeValue() {
            hello.value = "Hello Vue4";
            console.log(hello);
            obj.value.a=3;
            obj.value.b=4;
            //改变响应式数据 obj1 的属性 a 值为 9,reactive 响应式数据取值和赋值都不需
              要.value
            obj1.a=9;
            obj1.b=10;
            obj2[2]=5;                      //改变响应式数据 obj2 下标为 2 的数组元素值为 5
        }
        return {
            hello,
            obj,
            obj1,
            obj2,
            getHelloValue,
            changeValue,
        };
    },
};
</script>
```

启动项目,在浏览器地址栏输入访问地址 http://localhost:5173/,页面显示 Hello Vue3{"a":1,"b":2}{"a":7,"b":8}[1,2,3],然后单击"改变值"按钮,页面数据同步更新变为 Hello Vue4{"a":3,"b":4}{"a":9,"b":10}[1,2,5]。

2. 渲染响应式数据

在数据→页面方向,Vue 中数据值改变能够被页面 DOM 自动监听并更新渲染页面。

在上述的案例中,响应式数据的渲染统一采用插值表达式写法,插值表达式是一种最简单的渲染方式,直接在页面中显示数据。除此之外,还有一些较为复杂的渲染方式,这些渲染方式需结合 Vue 提供的指令实现。可以使用 Vue 提供的指令将响应式数据绑定到各种 HTML 标签的属性上,使得页面在不操作 DOM 元素的情况下根据响应式数据值的变化自动改变 DOM 元素的状态,如标签样式的变化、按钮状态的变化等。下面介绍渲染响应式数据时常用 Vue 的指令。

(1) v-bind。v-bind 指令主要用于单向渲染响应式数据值。单向渲染意味着只能实现数据→页面方向的渲染,数据改变页面随着渲染最新数据。v-bind 指令一般用于页面标签的属性渲染,用来绑定面标签的 class 属性或 style 样式中的属性。v-bind 指令支持绑定简单数据类型、对象类型和数组类型的响应式数据。v-bind 指令的使用语法如下。

```
<p v-bind:HTML 属性名="响应式数据变量名或表达式"></P>
```

如下是一些具体的使用示例。

```
<p v-bind:style="pStyle"></P>    //绑定响应式数据 pStyle 到<p>标签的 style 属性上
<p v-bind:class="pClass"></p>    //绑定响应式数据 pClass 到<p>标签的 class 属性上
```

其中,pStyle 和 pClass 可以是简单数据类型、对象类型和数组类型的响应式数据。使用时,v-bind 指令也支持缩写格式,上述两处代码可缩写为如下形式。

```
<p :style="style"></P>    //绑定响应式数据 style 的值到<p>标签的 style 属性上
<p :class="class"></P>    //绑定响应式数据 class 的值到<p>标签的 class 属性上
```

下面编程演示 v-bind 的使用。在 src/components 目录下新建 vbind.vue 文件,在内部添加如下代码。

【vbind.vue】

```
<template>
    <p v-bind:style="pStyle">{{pText}}</p>  //pStyle 绑定到<p>标签的 style 属性上
    <button @click="changeStyle">变绿</button>
</template>
<script>
import { reactive, ref } from 'vue';
export default {
    setup() {
        let pText=reactive("pText");
        let pStyle=reactive({color:"red"});
        function changeStyle() {
            pStyle.color = "green";             //改变<p>标签字体颜色为绿色
        }
        return {
            pText,
            pStyle,
            changeStyle,
        };
```

```
        },
    };
</script>
```

上述代码定义了 pText 和 pStyle 两个变量和一个函数 changeStyle，其中，pText 值为简单类型响应式数据，值在页面利用插值表达式显示。pStyle 是对象类型响应式数据，绑定在<p>标签的 style 属性上。如果单击"变绿"按钮，会执行 changeStyle 函数，将 pStyle 变量的 color 属性变为 green，pText 文本也会变成绿色。

在 mian.js 中使用如下代码，将 vbind.vue 挂载在 mian.js 中。

```
import vbind from './components/vbind.vue'
createApp(vbind).mount('#app')
```

启动项目，在浏览器中输入访问地址 http://localhost:5173/，页面文本 pText 初始颜色为红，单击"变绿"按钮，则 pText 颜色变为绿色，实现了样式的动态变化。

v-bind 除了绑定变量外，还可以绑定表达式的值。如实现功能，单击某个按钮让字体变绿，再单击该按钮字体变红，可对上述代码进行如下修改。

```
<template>
    <p v-bind:style="pStyle">{{pText}}</p>
    <!-- style 值为表达式,利用三目运算符进行条件判断,
    如果 flag 为 true,表达式为 color:red,否则为 color:green-->
    <p v-bind:style="flag==true?'color:red':'color:green'">变色</p>
    <button @click="changeStyle">切换颜色</button>
</template>
<script>
import { reactive, ref} from 'vue';
export default {
    setup() {
        let flag=ref(true);                    //定义 boolean 型响应式数据
        let pText=ref("pText");
        let pStyle=reactive({color:"red"});
        function changeStyle() {
            flag.value=!flag.value;            //每次单击按钮,flag 取反
            pStyle.color = "green";            //改变<p>标签字体颜色为样式
        }
        return {
            flag,
            pText,
            pStyle,
            changeStyle,
        };
    },
};
</script>
```

启动项目，在浏览器地址栏输入访问地址 http://localhost:5173/，变色字样初始颜色

为红,单击"切换颜色"按钮,变色字样颜色会在红绿之间反复变化。

(2) v-if 相关指令。v-if 相关指令包括 v-if、v-else-if 和 v-else,类似于 JS 中的 if、else if 和 else,用于渲染响应式数据时进行条件判断。其中,v-if 值为一个布尔型表达式,表达式返回真时渲染数据,反之不渲染。v-else-if 与最近的 v-if 匹配,当 v-if 中的表达式返回值为假时,才判断 v-else-if 的表达式值是否为真,表达式返回真时渲染数据,反之不渲染。条件判断中 v-else-if 可以有多个。v-else 与最近的 v-else-if 或 v-if 匹配,当 v-else-if 与 v-if 的表达式值均为假时,渲染 v-else 中的内容。v-if 相关指令的使用示例如下。

```
<template>
    <div v-if="条件表达式 A">条件表达式 A 为真输出此内容</div>
    <div v-else-if="条件表达式 B'">条件表达式 B 为真输出此内容</div>
    <div v-else-if=""条件表达式 C'">条件表达式 B 为真输出此内容</div>
    <div v-else>上述表达式都为假时输出此内容</div>
</template>
```

下面演示 v-if 相关指令用法,页面根据学生成绩等级不同,分别输出 A、B、C 和不及格。在 src/components 目录下新建 vif.vue 文件,内部输入如下内容。

【vif.vue】

```
<template>
    <div v-if="grade=== 'A'">A</div><!-- grade 值为 A,页面内容输出 A -->
    <div v-else-if="grade === 'B'">B</div> <!-- grade 值为 B,页面内容输出 B -->
    <div v-else-if="grade === 'C'">C</div>    <!-- grade 值为 C,页面内容输出 C-->
    <div v-else>不及格</div><!--否则页面内容为不及格-->
</template>
<script>
import {ref} from 'vue';
export default {
    setup() {
        let grade=ref("D");
        return {
            grade,
        };
    },
};
</script>
```

在 mian.js 中使用如下代码,将 vif.vue 挂载在 mian.js 中。

```
import vif from './components/vif.vue'
createApp(vif).mount('#app')
```

启动项目,在浏览器地址栏中输入访问地址 http://localhost:5173/,由于 grade 值为 D,因此,页面显示不及格。

(3) v-show。v-show 指令的值为一个布尔型变量,其作用是根据响应式数据值切换元素的显示状态。如果响应式数据值为 true,则显示元素内容,否则不显示,相当于给页面元素添加 display:none 样式。下面演示 v-show 指令用法,实现单击按钮,DOM 元素内容在隐藏和显示中来回切换。

在 src/components 目录下新建 vshow.vue 文件,内部输入如下内容。

【vshow.vue】

```
<template>
    <div v-show="show">11111</div>
    <button @click="showOrHidden">{{bText}}</button>
</template>
<script>
import { ref } from 'vue';
export default {
    setup() {
        let bText=ref("隐藏");
        let show=ref(true);
        function showOrHidden(){
            show.value=!show.value;         //变量 show 的值取反
            if(show.value){
                bText.value="隐藏";          //如果变量 show 的值为真,按钮文本为隐藏
            }else{
                bText.value="显示";          //如果变量 show 的值为假,按钮文本为显示
            }
        };
        return {
            bText,
            show,
            showOrHidden,
        };
    },
};
</script>
```

在 mian.js 中使用如下代码,将 vshow.vue 挂载在 mian.js 中。

```
import vshow from './components/vshow.vue'
createApp(vshow).mount('#app')
```

启动项目,在浏览器地址栏中输入访问地址 http://localhost:5173/,可以看到 11111 文本初始状态为显示,单击"隐藏"按钮,11111 文本在显示和隐藏之间来回切换,按钮内部的文本也随着变化。

(4) v-for。v-for 指令用于遍历一个数组或一个对象内部所有属性,在页面以列表形式渲染响应式数据内容。使用时 v-for 需结合 in 来使用,写法类似 for 循环中的 item in items 的形式。其中 items 是源数据数组或对象,而 item 则是被迭代的数组元素的别名或对象属性名。同时,v-for 还支持一个可选参数 index,即当前项的索引,遍历时索引从 0 开始计数。v-for 指令的使用语法如下。如数组名为 items,遍历数据 items 生成列表,代码写法如下。

```
<li v-for="(item,index) in items">
    {{ index }} - {{ item }}
</li>
```

下面演示 v-for 指令的用法,实现遍历对象数组生成 select 下拉框,同时遍历对象将内部所有属性打印生成列表。

在 src/components 目录下新建 vfor.vue 文件,内部输入如下内容。

【vfor.vue】

```
<template>
    <!-- 使用 v-for 遍历对象数组生成下拉框 -->
    <select>
        <!--此处使用:value 是为每个 option 添加 value 属性,属性值为 index 变量-->
        <option v-for="(item,index) in array" :value=index>{{item.id}}
</option>
    </select><br>
    <!-- 使用 v-for 遍历对象属性生成列表 -->
    <div style="grid: 0%;">
        <li v-for="(item,index) in obj">属性{{index}}:{{item}}</li>
    </div><br>
</template>
<script>
import { reactive } from 'vue';
export default {
    setup() {
        let array=reactive([{id:1},{id:2},{id:3}]);    //创建对象数组类型响应式数据
        let obj=reactive({id:1,name:'aaa',gender:'male'});//创建对象类型响应式数据
        return {
            array,
            obj,
        };
    },
};
</script>
```

在 mian.js 中使用如下代码,将 vfor.vue 挂载在 mian.js 中。

```
import vfor from './components/vfor.vue'
createApp(vfor).mount('#app')
```

启动项目,在浏览器地址栏中输入访问地址 http://localhost:5173/,可以看到页面生成了一个下拉框,内部有三个选项,分别是 1、2、3。生成的一个列表内容为"属性 id:1"、属性"name:aaa"、属性"gender:male"。

如果数组发生改变,Vue 会检测到数组变化,v-for 渲染的页面视图也会立即更新。Vue 将监听数组变更的方法进行了封装。改变数组时,只有使用这些方法才会触发页面视图更新。常用的变更数组方法如表 9-2 所示。

表 9-2 一些常用的变更数组方法

名 称	说 明
push()	将一个或多个元素添加至数组末尾,返回新数组的长度
pop()	从数组中删除最后一个元素并返回该元素
shift()	从数组中删除最后一个元素并返回该元素
unshift()	将一个或多个元素添加至数组开头,并返回新数组的长度
splice()	从数组中删除元素或向数组添加元素
sort()	对数组元素排序,默认按照 Unicode 编码排序,返回排序后的数组
reverse()	将数组中的元素位置颠倒,返回颠倒后的数组

此处需注意,当直接使用下标修改数组元素值,使用属性名修改对象属性值以及修改数组长度时,Vue 并不会将其加入响应式数据中,即使数据被修改,视图也不会进行更新。如 arr[索引]=新值、obj.属性名=新值、arr.length=1,这些都不能使视图更新数据。

如需改变对象,可以采用如下写法,页面会同步渲染。这里假设对象名为 obj。

```
obj.age=20;                    //向 obj 对象添加一个 age 属性赋值 20
obj.age=15;                    //将 obj 对象 age 属性修改为 15
delete obj.age;                //删除 obj 对象的 age 属性
```

下面演示如何修改数组和对象,使页面同步更新渲染。修改 vfor.vue 文件,在内部添加如下代码。

```
<template>
    <!-- 使用 v-for 遍历对象数组生成下拉框 -->
    <select>
        <option v-for="(item,index) in array" :value=index>{{item.id}}
</option>
    </select><br>
    <!-- 使用 v-for 遍历对象属性生成列表 -->
    <div style="grid: 0%;">
        <li v-for="(item,index) in obj">属性{{index}}:{{item}}</li>
    </div><br>
    <button @click="addArrayElement">添加数组元素和对象属性</button>
    <button @click="removeArrayElement">删除数组元素和对象属性</button>
    <button @click="reverseArray">颠倒数组元素</button>
</template>
<script>
import { reactive } from 'vue';
export default {
    setup() {
        let array=reactive([{id:1},{id:2},{id:3}]);
        let obj=reactive({id:1,name:'aaa',gender:'male'});
        function addArrayElement(){
            array.push({id:4});        //添加数据元素
            obj.age=20;                //添加对象属性
        };
```

```
        function removeArrayElement(){
            array.pop();                //删除数组最后一个元素
            delete obj.age;             //删除obj对象的age属性
        };
        function reverseArray(){
            array.reverse();            //颠倒数组
        };
        return {
            array,
            obj,
            addArrayElement,
            removeArrayElement,
            reverseArray,
        };
    },
};
</script>
```

启动项目,在浏览器地址栏输入访问地址 http://localhost:5173/,可以看到页面生成了一个下拉框,内部有三个下拉选项分别是 1、2、3。还生成了一个列表内容,属性"id:1"、属性"name: aaa"、属性"gender:male"。单击"添加数组元素和对象属性"按钮,页面下拉框新增了一个下拉选项 4,同时页面新增了属性"age:20"。单击"删除数组元素和对象属性"按钮,下拉选项 4 和属性"age:20"都消失了。单击"颠倒数组元素"按钮,下拉框中下拉选项被颠倒为 321 排列。

(5) v-for 和 v-if 结合使用。在使用 v-for 循环渲染数组数据时,往往需要对满足条件的部分数据进行特殊处理,如所有数据渲染完给 id 为 1 的数据添加高亮样式,所有数据渲染完默认选中 id=1 的数据等。这时就需要把 v-for 和 v-if 结合使用。使用时,尽量不要在同一元素上同时使用 v-if 和 v-for。由于 v-if 比 v-for 优先级更高,优先执行 v-if,因此 v-if 没有权限访问 v-for 里的变量,无法根据变量值进行条件判断。需将 v-for 写在外层的 <template>标签,内层才能使用 v-if 做条件判断。

下面演示 v-for 和 v-if 的结合使用,实现页面动态生成下拉框后默认选中其中某个下拉选项。修改 vfor.vue 文件,在内部添加如下代码。

```
<select>
    <template v-for="(item,index) in array">
        <option v-if="index==2" :value=index selected>{{item.id}}</option>
        <option v-else :value=index>{{item.id}}</option>
    </template>
</select><br>
```

上述代码将 v-for 写在外层的<template>标签内,在<template>标签内层利用 v-if 进行条件判断,如果添加的下拉选项 index 值为 2,在该下拉选项中添加 selected 属性,选中该下拉选项。

9.4.7 修改响应式数据

在 9.4.6 小节介绍了数据→页面方向的数据绑定和渲染,这种绑定是单向的。这一小节将介绍页面→数据方向的数据绑定,即用户操作页面进行一些数据修改操作,响应式数据值也能及时更新。这两个方向的数据绑定合并起来就是 Vue 的双向数据绑定。页面→数据方向的数据绑定多发生在表单类页面标签中,如输入框、下拉框、单选框和复选框等。这些组件值一方面需要根据响应式数据值的变化而变化,另一方面用户在页面的组件数据修改操作,也能被同步并更新响应式数据的值。对于一些非表单类标签如<p>、<h1>和<span>等,这些组件值不能被用户修改,只需要实现数据→页面方向的单向数据绑定。

要实现页面→数据方向的数据绑定,需要借助 Vue 中的 v-model 指令,该指令用于在<input>、<textarea>及<select>等表单页面标签上创建双向数据绑定。当页面加载时,各表单元素会从响应式数据中获取 value 属性值并渲染在页面上。如果用户更改了表单元素的值,v-model 会将各表单元素新的 value 属性值同步给响应式数据。使用时,直接在 HTML 表单类页面标签中添加 v-model="变量名就行",使用示例如下。

```
<input v-model="变量名">
<textarea v-model="变量名"></textarea>
<select v-model="变量名"></select>
```

下面以 input 输入框、单选框、复选框和下拉框为例,演示 v-model 用法。在 src/components 目录下新建 vmodel.vue 文件,内部输入如下代码创建输入框、单选框、复选框和下拉框。

【vmodel.vue】

```
<template>
<div style="grid: 0%;">
    <!-- 输入框 -->
   <input v-model="inputData">输入框值:{{inputData}} <br>
    <!-- 单选框 -->
   <input type="radio" value="male" v-model="radioData">male
   <input type="radio" value="female" v-model="radioData">female
单选框值:{{radioData}}<br>
    <!-- 动态生成复选框 -->
   <template v-for="(item,index) in array">
       <input type="checkbox" :value="item.id" v-model="ckDate.list">{{item.id}}
   </template> 复选框值:{{ckDate.list}}<br>
    <!-- 动态生成下拉框 -->
   <select name="select" id="select" v-model="selectOption">
      <template v-for="(item,index) in array">
         <option v-if="index==selectOption" :value=index selected>{{item.id}}</option>
         <option v-else :value=index>{{item.id}}</option>
      </template>
   </select>下拉框值:{{selectOption}}
</div>
</template>
<script>
import { ref,reactive } from 'vue';
```

```
export default {
    setup() {
        let inputData=ref("input");        //输入框数据
        let radioData=ref("male");         //单选框数据
        //复选框数据,数据→页面方向的数据绑定中,reactive 内的数据必须是对象或对象数
        //组,此处不能写成 reactive([])
        let ckDate=reactive({list:[]});
        let selectOption=ref(2);
        let array=reactive([{id:1},{id:2},{id:3}]);
        return {
            inputData,
            radioData,
            ckDate,
            selectOption,
            array
        };
    },
};
</script>
```

在 mian.js 中使用如下代码,将 vmodel.vue 挂载在 mian.js 中。

```
import vmodel from './components/vmodel.vue'
createApp(vmodel).mount('#app')
```

启动项目,在浏览器地址栏输入访问地址 http://localhost:5173/,可以看到页面生成了输入框、单选框、复选框和下拉框,手动修改任何一个表单元素,页面上都会同步显示最新的数据。

此外,v-model 还有三个修饰符可以使用,具体如下。

- lazy:一般用于输入类表单元素延迟数据同步时间。只在 input 输入框失去焦点才触发数据同步。使用示例如下:

```
v-model.lazy="变量名"
```

- trim:一般用于输入类表单元素去把用户输入字符串前后空格去掉。使用示例如下:

```
v-model.trim="变量名"
```

- number:一般用于输入类表单元素,将用户输入字符串转换成数值型,用户将不能输入非数字类字符。使用示例如下:

```
v-model.number="变量名"
```

9.4.8 异步加载响应式数据——Axios 组件

在前面的内容中,响应式数据的值都是人为设定的,但是在实际开发中,肯定会存在前

后端的数据交互,通常前端页面利用 ajax 发送异步请求,并加载后端的响应数据。Vue 中也存在有类似于 Ajax 的异步请求组件——Axios。Axios 是一个基于 Promise 的 HTTP 客户端,可以在浏览器和 node.js 中发送异步请求,因此 Vue 使用 Axios 组件来完成异步请求功能。

1. 安装 Axios

使用 Axios 之前需先在 Vue 项目中安装 Axios。安装 Axios 的步骤非常简单,这里在 vue001 项目中安装 Axios,打开 VsCode 命令行终端,进入 vue001 项目的根目录,输入 npm install axios 命令即可安装,安装结果如图 9-19 所示。

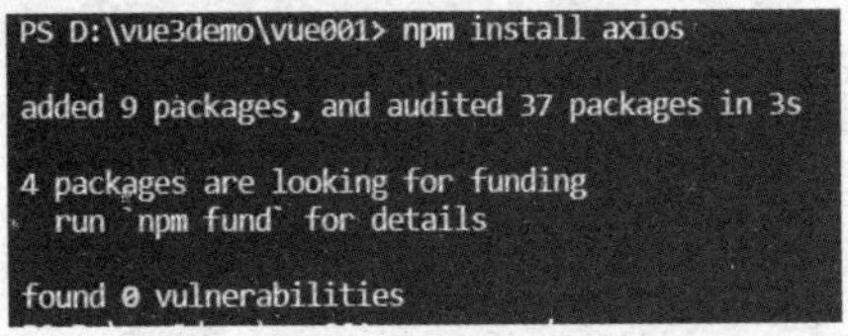

图 9-19 安装 Axios

安装完毕,单击 package.json 文件,在 package.json 文件内部可以看到如下信息,Axios 的版本为 1.4.0。

```
"dependencies": {
    "axios": "^1.4.0",
    "vue": "^3.3.2"
},
```

只需在 Vue 组件的＜script＞标签中输入如下命令,局部引入 Axios 即可直接使用,当然也可以全局配置引入。为方便,这里采用局部引入 Axios。

```
import axios from 'axios'
```

2. Axios 的使用语法

下面介绍 Axios 的使用语法,Axios 请求主要分为 get 和 post 请求。其中,get 请求方式写法如下。

```
axios.get('请求 URL')
    .then(response => {
        //请求成功,响应数据使用 response.data 获取
        console.log(response.data)
    }).catch(error => {
        //处理请求失败
        console.log(error)
    });
```

get 方式如需传递请求参数,可采用如下三种写法。

```
//传统请求传参方式,直接在请求 URL 后添加请求参数 id=123
axios.get('/user?id=123')
```

```
//Restful 风格请求,在请求 URL 内部添加请求参数 123
axios.get('/user/123')
//通过 params 设置请求参数
axios.get('/user', {
    params: {
      id:123
    }
})
```

如果发送的是 post 请求,使用语法和 get 类似。

```
axios.post('请求 URL')
    .then(response => {
      请求成功,响应数据使用 response.data 获取
      console.log(response.data)
    }).catch(error => {
      //处理请求失败
      console.log(error)
});
```

post 方式如需传递请求参数,可采用如下写法。

```
axios.post('请求 URL', {
    id: 123,                              //参数 id
    name: 'zhangsan'                      //参数 name
})
```

post 方式传递 Json 对象可以可采用如下写法。这里假设传递 User 对象,内部包含 username 和 password 两个属性。

```
axios({
    url:'请求路径',
    data: {
        username:'zhangsan',
        password:'123456'
    },
    method:'POST',
    headers:{
        'Content-Type':'application/json'
    }
})
```

3. Axios 跨域问题

在前后端分离的项目中,使用 Axios 向后端接口异步请求数据必然会产生跨域问题,导致请求失败。跨域产生的原因是浏览器访问的前端网页与前端网页中 Axios 向后端请求数据的路径不在同一个域或同一个端口,从而违反了浏览器访问资源的规则。如在浏览器输入 http://localhost:5173/访问 Vue 页面时,如果该页面挂载的 Vue 组件中使用 Axios 向后端接口异步请求数据。而后端是一个独立的项目,接口路径往往和 http://localhost:

5173/不在同一个域或同一个端口,这是浏览器访问所不允许的。

如果没有配置跨域,Axios 请求访问后端接口时,浏览器控制台会打印如下失败信息,这里假设请求的后端地址为 http://127.0.0.1:9999。

```
Access to XMLHttpRequest at 'http://127.0.0.1:9999/' from origin 'http://localhost:5173/'
has been blocked by CORS policy: No 'Access-Control-Allow-Origin' header is present on the requested resource.
```

在 Vue3 项目中解决跨域问题非常简单,只需在项目的 vite.config.js 文件中添加如下配置。这里假设后端地址为 http://127.0.0.1:9999。

```
import { fileURLToPath, URL } from 'node:url'
import { defineConfig } from 'vite'
import vue from '@vitejs/plugin-vue'
//https://vitejs.dev/config/
export default defineConfig({
  plugins: [vue()],
  resolve: {
    alias: {
      '@': fileURLToPath(new URL('./src', import.meta.url))
    }
  },
  server: {
    open: true,
    host: 'localhost',              //前端项目的 IP 地址
    port: 5173,                     //前端项目的端口号
    proxy: {
      //添加一个代理,将/api 开头的请求地址映射为 http://127.0.0.1:9999
      '/api': {
        target: 'http://127.0.0.1:9999',   //后端项目的访问地址
        ws: true,
        changeOrigin: true,
        //替换 target 中的后端项目的访问地址,将 Axios 中的请求地址 http://127.0.0.1:
          //9999/xxx 替换为/api/xxx
        rewrite: (path) => path.replace(/^\/api/, '')
      },
      //此处可以根据需求添加多个代理,分别映射不同的后端请求地址,
      //实现 Vue 项目和多个不同服务端进行数据交互
    }
  }
})
```

4. Axios 编程应用

下面在 vue001 项目中添加跨域配置来演示 Axios 的编程应用。其中,Axios 请求的数据来自 runoob 提供的开源接口,接口数据结构如图 9-20 所示。页面需以列表显示图 9-20 所示 sites 数组内三个元素 name 和 info 属性的内容。

在 vue001 项目的 vite.config.js 文件中添加访问 runoob 开源接口的跨域配置。

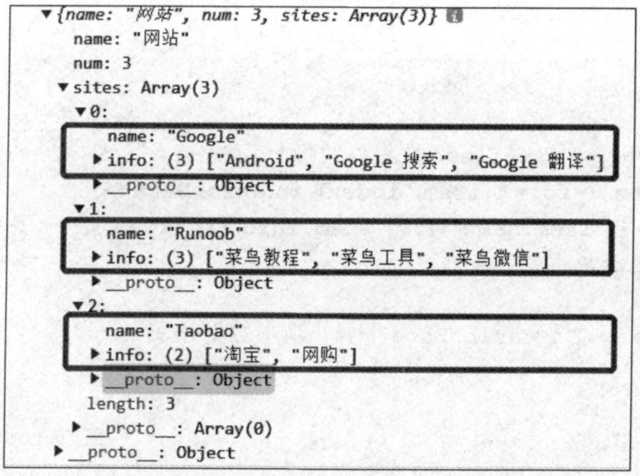

图 9-20　runoob 开源接口数据

【vite.config.js】

```
import { fileURLToPath, URL } from 'node:url'
import { defineConfig } from 'vite'
import vue from '@vitejs/plugin-vue'
//https://vitejs.dev/config/
export default defineConfig({
  plugins: [vue()],
  resolve: {
    alias: {
      '@': fileURLToPath(new URL('./src', import.meta.url))
    }
  },
  server: {
    open: true,
    host: 'localhost',            //前端项目的 IP 地址
    port: 5173,                   //前端项目的端口号
    proxy: {
      //添加一个代理,将/api 开头的请求地址映射为 target 内的地址
      '/api': {
        target: 'https://www.runoob.com',//后端项目的访问地址
        ws: true,
        changeOrigin: true,
        //替换 target 中的请求地址,将 Axios 中的请求地址 https://www.runoob.com/xxx
        //替换为/api/xxx
        rewrite: (path) => path.replace(/^\/api/, '')
      },
    }
  }
})
```

在 src/components 目录下新建 axios.vue 文件,内部输入如下代码访问 runoob 开源接口数据。

【axios.vue】

```vue
<template>
    <div style="grid: 0%;">
        <!-- 循环打印 axiosData 数组内的每个元素 -->
        <template v-for="(item, index) in axiosData">
            <li>{{ item.name }},{{ item.info }}</li>
        </template>
    </div><br>
    <button @click="getaxiosData">异步请求数据</button>
</template>
<script>
import { reactive} from 'vue';
import axios from 'axios'                       //引入 axios 组件
export default {
    setup() {
        let axiosData = reactive([]);           //reactive 创建数组类型响应式数据
        function getaxiosData() {
            //每次请求之前先清空 axiosData 数组的元素,否则每次单击按钮都会新增三个列
            //表元素
            axiosData.splice(0, axiosData.length);
            axios.get('/api/try/ajax/json_demo.json')   //请求路径以/api 开头
                .then(response => {
                    console.log(response.data);
                    let resData = response.data.sites;
                    //将收到的数据依次推进 axiosData 数组
                    for (let obj in resData) {
                        axiosdata.push(resData[obj]);
                    }
                }).catch(error => {
                    //打印请求失败信息
                    console.log(error)
                });
        };
        return {
            axiosData,
            getaxiosData,
        };
    },
};
</script>
```

在 mian.js 中使用如下代码,将 axios.vue 挂载在 mian.js 中。

```
import axios from './components/axios.vue'
createApp(axios).mount('#app')
```

启动项目,在浏览器地址栏访问地址 http://localhost:5173/,单击页面中的"异步请求数据"按钮,页面上成功显示 runoob 开源接口响应的数据,如图 9-21 所示。

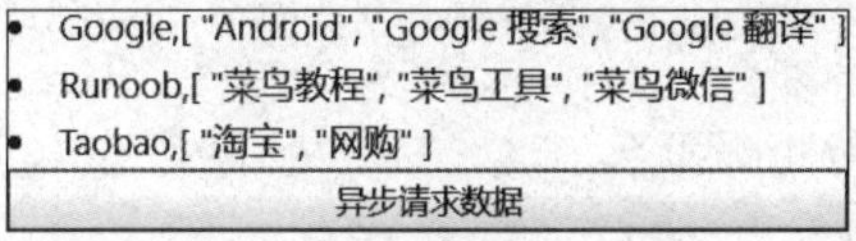

- Google,["Android", "Google 搜索", "Google 翻译"]
- Runoob,["菜鸟教程", "菜鸟工具", "菜鸟微信"]
- Taobao,["淘宝", "网购"]

异步请求数据

图 9-21　显示 runoob 开源接口响应的数据

9.4.9　Vue3 页面跳转——Vue-Router 组件

在传统的 Web 项目中，前后端不分离，页面跳转依赖于项目后端 Servlet 实现，Servlet 根据请求 URL 判断跳转到哪个页面。而在前后端分离的项目中，页面跳转不再依赖于后端实现。前端项目是一个独立的项目，由自己管理前端页面的跳转。在 Vue3 中，实现前端页面的跳转需依赖于 Vue 的路由组件 Vue-Router，版本信息为 Vue-Router4 版本。Vue 路由组件允许我们通过不同的 URL 访问不同的 Vue 组件。实现多视图的单页面应用。

1. 安装 Vue-Router

使用 Vue-Router 之前需先在 Vue 项目中安装 Vue-Router。安装 Vue-Router 步骤非常简单，这里在 vue001 项目中安装 Vue-Router，打开 VsCode 命令行终端，进入 vue001 项目的根目录，输入 npm install vue-router@4 命令即可安装，安装结果如图 9-22 所示。

```
PS D:\vue3demo\vue001> npm install vue-router@4

added 2 packages, and audited 28 packages in 16s

3 packages are looking for funding
  run `npm fund` for details

found 0 vulnerabilities
PS D:\vue3demo\vue001>
```

图 9-22　安装 Vue-Router

安装完毕，在 package.json 文件内部可以看到如下信息，Vue-Router 的版本为 4.2.2。

```
"dependencies": {
    "axios": "^1.4.0",
    "vue": "^3.3.2",
    "vue-router": "^4.2.2"
},
```

2. 配置 Vue-Router 路由

Vue-Router 路由组件一般采用全局配置引入，将所有的路由映射以列表形式统一配置在一个文件中，将该路由映射列表挂载在 Vue 实例上引入。具体实现如下：在 Vue 项目的 src 目录下新建一个 router 文件夹，在 router 文件夹内部新建 index.js 文件，在 index.js 文件内配置全局路由映射。

```
import {createRouter, createWebHashHistory} from 'vue-router';
//定义路由映射列表,内部可以定义多个路由映射
const routes = [
    {
        path: '/url1',                    //访问当前的 URL。'/'表示 vue 项目根目录
        name: 'url1',                     //路由命名
```

```
        component: () => import(跳转到的Vue组件)
    },
    //内部可以添加多个路由映射
]
//导出路由
export default createRouter({
    history: createWebHashHistory(),
    routes
})
```

在 Vue 项目的 main.js 文件中,全局导入 index.js 内的路由配置,将 router 全局挂载在创建的 Vue 实例上,这样就可以在自己的 Vue 组件中使用 Vue-Router 了。

```
import './assets/main.css'
import { createApp } from 'vue'
import App from './App.vue'
import router from './router/index'          //导入路由配置文件 index.js
createApp(App).use(router).mount('#app')  //将 router 全局挂载在创建的 Vue 实例上
```

3. Vue-Router 使用语法

Vue-Router 利用 useRouter 函数的 push 方法实现页面跳转,匹配路由映射的组件使用＜router-view＞标签接收,并作为子组件渲染在＜router-view＞标签所在的 Vue 组件上。使用 Vue 开发单页面应用中,一般将＜router-view＞标签写在 App.vue 上,以 App.vue 作为父组件。在 index.js 中统一配置不同 Vue 组件的访问 URL,就可以通过路由匹配将不同的 Vue 组件作为子组件渲染在 App.vue 上,实现页面跳转。

useRouter 函数使用示例如下。在＜script＞标签从 Vue-Router 中引入 useRouter 函数,并调用 push 方法。

```
<script>
import {useRouter} from 'vue-router';
export default {
    setup() {
        const router = useRouter();
        router.push({
            path:'/aaa',
        });
    },
};
</script>
```

上述代码将在当前页面加载 URL 为/aaa 所映射的 Vue 组件。如在页面跳转时需要传递参数,可使用 query 和 params 两种方式进行传参。其中,query 方式传参拼接在 URL 后面,可使用路由的 path 属性匹配 URL。params 方式传递的参数必须作为 URL 的一部分,路由路径上必须有":参数名"(此处为 Vue-Router4 新做的修改,老版本没有要求),只能使用路由的 name 属性匹配 URL,不能使用 path 属性匹配。下面分别介绍 query 传参和 params 传参的使用语法。

(1) query 传参。假设父组件传递参数名为 queryData 且值为 data 变量的值,页面跳转的子组件为 src/components 目录下的 queryData.vue。query 传参的使用语法如下。

```
<template>
    <RouterView></RouterView> <!-- 路由匹配到的组件将渲染在这里 -->
</template>
<script>
import { useRoute } from "vue-router";
export default {
  setup() {
      //query 方式传递参数 data
      router.push({
          path:'/queryData,              //使用 path 匹配 URL
          query:{
              queryData:data             //传递参数名为 queryData
          }
      } );
  },
};
</script>
```

在 index.js 中需定义如下 URL 映射,并在 main.js 文件中引入挂载在创建的 Vue 实例上。

```
const routes = [
    {
        path: '/queryData',
        name: 'queryData',
        component: () => import('../components/queryData.vue')
    },
]
```

子组件 querydata.vue 在＜script＞标签从 Vue-Router 中引入 useRoute 函数,使用 useRoute 函数的 query 属性接收。此处引入 useRoute 组件,需和数据发送方使用的 useRouter 函数区分开。

```
<template>
  <span>query 方式接收数据:{{querydata}}</span>
</template>
<script>
import { useRoute } from "vue-router";
export default {
    setup() {
        const router = useRoute();
        let queryData=router.query.queryData;     //从参数名中获取传递的数据
        return {
            queryData,
        };
    },
```

253

```
};
</script>
```

(2) params 传参。假设传递参数名为 paramsData 且值为 Data 变量的值,页面跳转的子组件为 src/components 目录下的 paramsData.vue。params 传参的使用语法如下。

```
<template>
    <RouterView></RouterView> <!-- 路由匹配到的组件将渲染在这里 -->
</template>
<script>
import { useRoute } from "vue-router";
export default {
    setup() {
        //params 方式传递参数 data
        router.push({
            //这里只能使用路由命名 name 匹配,不能使用 path 匹配
            name: 'paramsHome',
            params:{
                paramsData:data,          //传递参数名为 paramsData
            }
        });
    },
};
</script>
```

上述 params 传参只能使用路由命名 name 匹配 URL,不能使用 path 匹配。

在 index.js 中需定义如下 URL 映射,这里需将参数名拼接在 path 路径内部,并在 main.js 文件中引入挂载在创建的 Vue 实例上。

```
const routes = [
    {
        path: '/paramsHome/:paramsData',      //这里需将参数名拼接在 path 路径内部
        name: 'paramsHome',
        component: () => import('../components/paramsData.Vue')
    },
]
```

子组件 paramsData.vue 在<script>标签从 Vue-Router 中引入 useRoute 函数,使用 useRoute 函数的 params 属性接收。此处引入 useRoute 组件,需和数据发送方使用的 useRouter 函数区分开。

```
<template>
  <span>params 方式接收参数:{{paramsData}}</span>
</template>
<script>
import { useRoute } from "vue-router";
export default {
```

```
    setup() {
        const router = useRoute();
        let paramsdata=router.params.paramsData;
        return {
            paramsData
        };
    },
};
</script>
```

除使用 useRouter 函数的 push 方法实现页面跳转外，Vue-Router 还封装了一个组件＜router-link＞用于页面跳转。＜router-link＞作用类似于＜a＞标签。＜router-link＞内部可以定义 to 属性，设置一个 URL 跳转页面。router-link 使用语法如下。

```
<router-link :to="/url1">page1</router-link>
<!-- 路由匹配到的组件将渲染在这里 -->
<router-view></router-view>
```

上述代码定义了一个＜router-link＞组件，并配置路由为/url1。单击＜router-link＞组件，＜router-link＞内部会把 to 属性的 URL 值传给 useRouter 函数，利用 useRouter 函数的 push 方法跳转页面。如需传递参数，也可以使用 params 和 query 两种传参方式，使用语法如下。

```
//query 方式传递参数 data
<router-link :to="{path:'/url1,query:{data}'}">query</router-link>
//params 方式传递参数 data
<router-link :to="{path:'/url1,params:{data}'}">params</router-link>
```

4. Vue-Router 编程应用

下面在 vue001 项目中演示 Vue-Router 的编程应用。实现功能：访问 Vue 项目根目录页面，自动跳转到登录页面 Login.vue。在登录页面 Login.vue 中输入用户名和密码，单击"登录"按钮。如果输入用户名和密码都为 admin，则登录成功，页面跳转到 Home 主页并显示登录用户名和密码，否则不跳转页面并显示登录失败信息。参数分别使用 query 和 params 两种方式实现传参。

将 vue001 项目的 App.vue 代码修改如下：清空＜template＞标签内容，并在＜template＞标签内重新添加＜RouterView＞＜/RouterView＞标签。App.vue 作为所有 Vue 组件的父组件，如果路由匹配，则在 App.vue 内渲染子组件。

```
<template>
    <RouterView></RouterView>
</template>
```

在 vue001 项目的 src 目录下新建一个 router 文件夹，在 router 文件夹内部新建 index.js 文件，在 index.js 文件内配置全局路由映射。

【index.js】

```js
import { createRouter, createWebHashHistory } from 'vue-router';
//定义路由列表
const routes = [
    {
        //访问项目根目录,跳转到 Login.vue
        path: '/',
        name: 'login',                      //路由命名
        component: () => import('../components/Login.vue')
    },
    {
        //登录成功,使用 query 方式传参,跳转到 QueryHome.vue
        path: '/queryHome/',
        name: 'queryhome',
        component: () => import('../components/QueryHome.vue')
    },
    {
        //登录成功,使用 params 方式传参,跳转到 ParamsHome.vue
        //此处 path 路径内部需添加传递的参数名 paramsdata
        path: '/paramsHome/:paramsData',
        name: 'paramshome',
        component: () => import('../components/ParamsHome.vue')
    }
]
//导出路由
export default createRouter({
    history: createWebHashHistory(),
    routes
})
```

修改 vue001 项目的 main.js 文件代码,内部导入 index.js 配置的路由列表,并挂载在 App.vue 上。

```js
import './assets/main.css'
import { createApp } from 'vue'
import App from './App.vue'
//导入新建的路由文件
import router from './router/index'
createApp(App).use(router).mount('#app')
```

在 vue001 项目的 src/components 目录下新建登录页面 Login.vue,在内部添加如下代码。

【Login.vue】

```html
<template>
    <div style="grid: 0%;">
        <span v-show="msg!=''" style="color: red;">{{ msg }}</span><br>
```

```html
            用户名:<input v-model="loginData.username">{{ loginData.username }}
            密码:<input v-model="loginData.password">{{ loginData.password }}<br>
            <button @click="loginQuery">query 传参登录</button>
            <button @click="loginParams">params 传参登录</button>
            <RouterView></RouterView>
    </div>
</template>
<script>
import { reactive, ref } from 'vue';
import { useRouter } from 'vue-router';
export default {
    setup() {
        //以对象类型响应式数据封装用户名和密码
        let loginData = reactive({
            username: "",
            password: ""
        });
        let msg=ref("");          //登录失败提示信息
        const router = useRouter();
        function loginQuery() {
            if (loginData.username == 'admin' && loginData.password == "admin") {
            //query 方式传递对象,需利用 JSON.stringify 将对象转为 JSON 字符串传递
                router.push({
                    path: '/queryHome',
                    query: {
                        queryData: JSON.stringify(logindata),
                    }
                });
            }else{
                msg.value="用户名密码错误,登录失败";
            }
        };
        function loginParams() {
          //params 方式必须使用 name 匹配路由,不能使用 path
          //params 参数属于路径当中的一部分,需在配置路由的时候,
          //在路径后面添加参数名,如 path: '/paramsHome/:list'
          //params 方式传递对象,需利用 JSON.stringify 将对象转为 JSON 字符串传递
            if (loginData.username == 'admin' && loginData.password == "admin") {
                router.push({
                    name: 'paramshome',
                    params: {
                        paramsData: JSON.stringify(loginData),
                    }
                });
            }else{
                msg.value="用户名密码错误,登录失败";
            }
        };
        return {
            loginData,
```

```
                msg,
                loginQuery,
                loginParams
            };
        },
    };
</script>
```

在上述代码中定义了 loginData 存储响应式对象类型数据,对象内部封装 username 和 password。定义了 msg 变量存储登录失败信息。页面添加了两个按钮,分别用于登录成功后 query 和 params 传参,如果用户登录成功,则分别跳转到 QueryHome.vue 和 ParamsHome.vue。否则提示登录失败,不跳转页面。由于 loginData 为对象类型数据,父组件发送数据时需利用 JSON.stringify 方法将对象格式化成 JSON 字符串传递,在子组件接收数据时又利用 JSON.parse 方法将 JSON 字符串转为 JSON 对象。

在 vue001 项目的 src/components 目录下新建 QueryHome.vue 和 ParamsHome.vue,在内部分别添加如下代码并获取登录用户名和密码。

【QueryHome.vue】

```
<template>
  <span>query 方式接收的用户名:{{queryDataObj.username}}</span>
  <span>query 方式接收的密码:{{queryDataObj.password}}</span>
</template>
<script>
import { useRoute } from "vue-router";
export default {
    setup() {
        const router = useRoute();
        let queryData=router.query.queryData;          //接收 queryData 参数数据
        let queryDataObj=JSON.parse(queryData);        //JSON 字符串转 JSON 对象
        return {
            queryDataObj,
        };
    },
};
</script>
```

【ParamsHome.vue】

```
<template>
  <span>params 方式接收的用户名:{{paramsDataObj.username}}</span>
  <span>params 方式接收的密码:{{paramsDataObj.password}}</span>
</template>
<script>
import { useRoute } from "vue-router";
export default {
```

```
    setup() {
        const router = useRoute();
        let obj=router.params.paramsData;          //接收 paramsData 参数数据
        let paramsDataObj=JSON.parse(obj);         //JSON 字符串转 JSON 对象
        return {
            paramsDataObj
        };
    },
};
</script>
```

启动项目,在浏览器访问地址 http://localhost:5173/,访问地址为根目录,页面跳转到 Login.vue,如图 9-23 所示。

图 9-23　Login.vue 页面

如果输入用户名和密码都不是 admin,单击两个"登录"按钮,页面都不跳转,显示登录失败信息,如图 9-24 所示。

图 9-24　登录失败

如果输入用户名和密码都是 admin,单击"query 传参登录"按钮,页面跳转到 QueryHome.vue,并显示登录的用户名和密码,如图 9-25 所示。

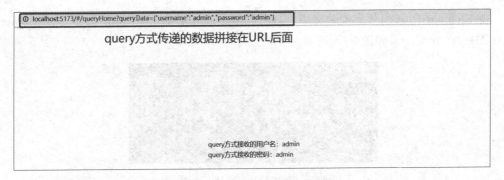

图 9-25　query 方式接收数据

如果输入的用户名和密码都是 admin,单击"params 传参登录"按钮,页面跳转到 ParamsHome.vue,并显示登录的用户名和密码,如图 9-26 所示。

9.4.10　Vue3 集成 Element-Plus

在开发 Vue 项目中,不可避免要自己编写一些 Vue 组件,组件内部根据需要会封装一些 UI 元素。为了使组件布局合理、样式精美,往往需要编写大量的 CSS 样式,这非常不方

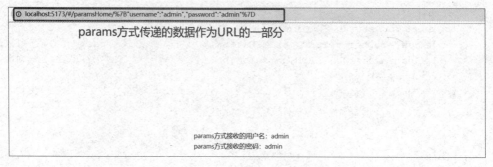

图 9-26 params 方式接收数据

便。Vue3 官方推荐使用 Element-Plus 构建页面 UI 元素，Element-Plus 是基于 Vue3 实现的一套纯粹的 UI 组件库，内部实现了很多现成的布局容器和 UI 元素，降低了用户编写 Vue 组件的难度。

1. 安装 Element-Plus

使用 Element-Plus 之前需先在 Vue 项目中安装 Element-Plus。安装 Element-Plus 分为两部分，需要分别安装 Element-Plus 组件库和图标库。这里在 vue001 项目中安装 Element-Plus，在 VsCode 命令行终端中进入 vue001 项目的根目录，输入 npm install element-plus 命令，安装 Element-Plus 组件库，安装结果如图 9-27 所示。

图 9-27 安装 Element-Plus 组件库

待 Element-Plus 组件库安装完毕，再输入 npm install element-plus 命令安装 Element-Plus 图标库，安装结果如图 9-28 所示。

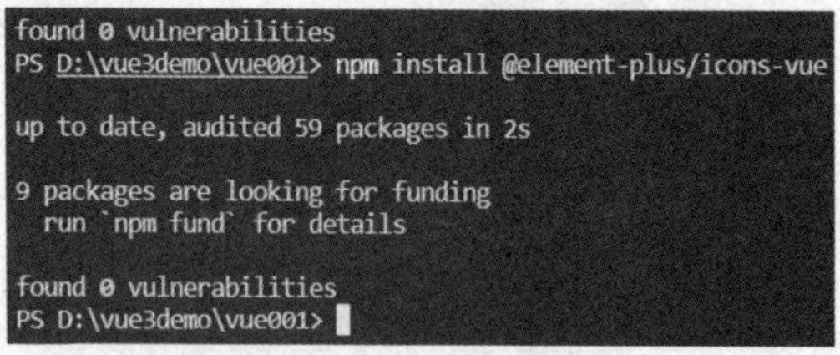

图 9-28 安装 Element-Plus 图标库

待 Element-Plus 图标库安装完毕，在 package.json 文件内部可以看到如下信息，Element-Plus 组件库的版本为 2.3.6，图标库版本为 2.1.0。

```
"dependencies": {
    "@element-plus/icons-vue": "^2.1.0",
    "axios": "^1.4.0",
    "element-plus": "^2.3.6",
```

```
    "vue": "^3.3.2",
    "vue-router": "^4.2.2"
},
```

2. 配置 Element-Plus

Element-Plus 一般采用全局配置引入,也可以局部配置引入,这里介绍全局配置引入。在 vue001 项目的 main.js 文件中添加如下代码,全局引入 Element-Plus 图标库和组件库。

```
import './assets/main.css'
import { createApp } from 'vue'
import App from './App.vue'
//导入新建的路由文件
import router from './router/index'
//导入 element-plus 组件库
import ElementPlus from 'element-plus'
//导入 element-plus 组件库样式
import 'element-plus/dist/index.css'
//导入 element-plus 图标库
import * as ElementPlusIconsVue from '@element-plus/icons-vue'
const app = createApp(App)
//全局挂载 element-plus 组件库在 Vue 实例上
app.use(router).use(ElementPlus).mount('#app')
//全局挂载 element-plus 图标库在 Vue 实例上
for (let iconName in ElementPlusIconsVue) {
    app.component(iconName, ElementPlusIconsVue[iconName])
}
```

上述代码中引入 Element-Plus 组件库、Element-Plus 组件库样式和 Element-Plus 图标库,并使用 Vue 实例的 use 函数全局挂载 ElementPlus 组件库,同时使用 for 循环将图标库中每个图标依次挂载在 Vue 实例上。下面就可以在页面中使用 Element-Plus 了。

3. 使用 Element-Plus

Element-Plus 中相关元素的标签统一以 el-前缀开头,即在原有 HTML 标签的基础上添加 el-。如按钮组件标签为<el-button>,输入框组件为<el-input>等。Element-Plus 官网提供了大量组件的样例代码,开发人员可以直接复制样例代码到 Vue 组件中使用。下面以一个表单案例演示 Element-Plus 的使用。

在 vue001 项目的 src/components 目录下新建 El_Form.vue 文件,在内部利用 Element-Plus 实现表单注册页面,内部 HTML 元素包括输入框、下拉框、单选框、复选框和文本域。

【El_Form.vue】

```
<template>
    <!--每个 form 组件中都需要 form-item 字段作为容器,用于获取值与验证值-->
    <el-form label-width="120px">
        <el-form-item label="用户名">
            <el-input v-model="regData.username" />
```

```html
            </el-form-item>
            <el-form-item label="密码">
                <el-input v-model="regData.password" />
            </el-form-item>
            <el-form-item label="性别">
                <el-select v-model="regData.gender" placeholder="select your gender">
                    <el-option label="男" value="0" />
                    <el-option label="女" value="1" />
                </el-select>
            </el-form-item>
            <el-form-item label="职业">
                <el-radio-group v-model="regData.radio">
                    <el-radio label="学生" />
                    <el-radio label="职员" />
                </el-radio-group>
            </el-form-item>
            <el-form-item label="爱好">
                <el-checkbox-group v-model="regData.ck">
                    <el-checkbox label="篮球" name="type" />
                    <el-checkbox label="足球" name="type" />
                </el-checkbox-group>
            </el-form-item>
            <el-form-item label="个人描述">
                <el-input v-model="regData.desc" type="textarea" />
            </el-form-item>
            <el-form-item>
                <el-button type="primary" @click="reg">提交</el-button>
                <el-button>取消</el-button>
            </el-form-item>
        </el-form>
</template>
<script>
import { reactive} from 'vue';
export default {
    setup() {
        //以对象类型响应式数据封装注册数据
        let regData = reactive({
            username: "",
            password: "",
            gender:"",
            radio:"",
            ck:[],
            desc:""
        });
        function reg() {
          console.log(regData);         //控制台打印输出 regData
        };
        return {
            regData,
            reg
```

```
        };
    },
};
</script>
```

修改 mian.js 中代码如下，将 El_Form.vue 挂载在 mian.js 中。

```
import './assets/main.css'
import { createApp } from 'vue'
import App from './App.vue'
import elform from './components/El_Form.vue'
//导入新建的路由文件
import router from './router/index'
//导入 element-plus 组件库
import ElementPlus from 'element-plus'
//导入 element-plus 组件库样式
import 'element-plus/dist/index.css'
//导入 element-plus 图标库
import * as ElementPlusIconsVue from '@element-plus/icons-vue'
const app = createApp(elform)
//全局挂载 element-plus 组件库到 Vue 实例上
app.use(router).use(ElementPlus).mount('#app')
//全局挂载 element-plus 图标库到 Vue 实例上
for (let iconName in ElementPlusIconsVue) {
    app.component(iconName, ElementPlusIconsVue[iconName])
}
```

启动项目，在浏览器地址栏中输入访问地址 http://localhost:5173/，页面显示 El_Form.vue 的内容，如图 9-29 所示，表单页面布局整齐，内部各 HTML 元素都使用了 Element-Plus 中的样式。

在表单中输入数据，单击"提交"按钮，控制台打印出变量 regdata 的值，如图 9-30 所示，表单内所有 HTML 元素的输入值都正常获取了。

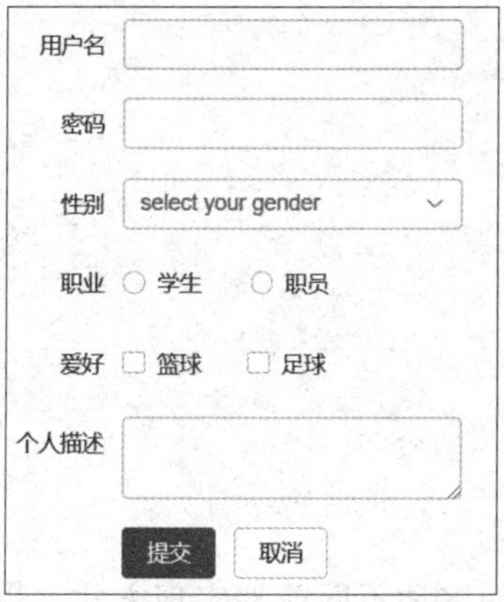

图 9-29　Element-Plus 表单

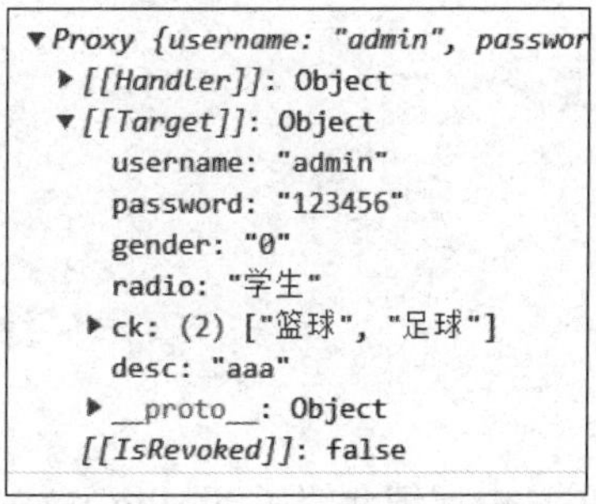

图 9-30　控制台打印用户输入的数据

任务 9.5　综合案例：基于 Vue3 实现员工信息管理

下面通过一个综合案例，演示在前后端分离架构下如何利用 Vue 构建前端页面并和后端 API 接口进行数据交互。

9.5.1　案例任务

在本项目任务 9.3 的 SpringDemo9_Thymeleaf 项目基础上，基于 MySQL 数据库中的数据库表 employee，employee 表字段为 emp_id、emp_name、emp_gender、emp_dept 和 emp_birth，并利用 Vue3＋Element-Plus 编程实现员工信息的增、删、改、查。

9.5.2　案例分析

该案例是一个前后端分离开发的综合案例，后端可基于原有 SpringDemo9_Thymeleaf 项目实现。前端基于 Vue3 实现，案例涉及任务 9.4 的内容包括 Vue 的数据绑定、路由跳转、数据异步加载和 Element-Plus 页面组件的编程等。案例编程时需考虑如下因素。

（1）由于项目前后端分离，后端控制器内部请求处理方法变成纯粹的 API 接口，不再涉及页面跳转，页面跳转由 Vue 负责。因此，原 SpringDemo9_Thymeleaf 项目的 EmployeeController 控制器内部方法不再使用，需重新定义员工信息的增、删、改、查 API 接口。

（2）Vue 项目中涉及数据异步加载和页面跳转，需进行路由和跨域设置。

（3）员工信息管理包含三个页面，分别是新增员工、修改员工和查询员工。其中，新增和修改员工页面需使用 Element-Plus 表单组件。查询员工信息需使用 Element-Plus 表格组件。需掌握这两个组件的详细使用方法。

9.5.3　任务实施

由于篇幅有限，任务实施具体步骤采用二维码形式展示。

综合案例的实施步骤

小　　结

本项目中详细介绍了 Spring Boot Web 应用开发的多种不同前端实现方式。以项目 8 中任务 8.9 的员工信息管理案例为例，分别利用 JSP、Thymeleaf 和 Vue 三种方式实现前端页面。章节知识点包括 JSP 和 Thymeleaf 模板引擎的语法和编程应用，Vue3 项目的创建及相关语法指令的使用，响应式数据的定义和使用，Vue3 项目集成及使用 Vue-Router、Axios 和 Element-Plus 组件等内容。最后以一个综合案例——基于 Vue3 实现员工信息管理，演示如何使用前后端分离开发模式开发基于 Spring Boot＋Vue 的 Web 应用。

课后练习：学生信息管理

模仿任务 9.5，在 MySQL 数据库中创建数据库表 student，student 表字段为 stu_id、stu_name、stu_gender、stu_class 和 stu_birth，并利用 Spring Boot＋Vue 前后端分离模式实现学生信息的增、删、改、查。

项目 10　Spring Boot 安全控制
——Security

在 Web 开发中,资源访问的安全控制功能属于非功能需求,但又是项目中必不可少的。在 Spring Boot 项目中可以利用 Spring Security 实现安全控制。Spring Security 是 Spring 生态的成员之一,能够和 Spring Boot 更好地集成在一起实现资源访问的安全控制功能。Spring Boot 能够自动对 Spring Security 进行默认配置,使开发人员更方便地使用 Spring Security。本项目将详细介绍如何在 Spring Boot 项目中使用 Spring Security 进行安全控制。

任务 10.1　初识 Spring Security

Spring Security 是 Web 开发中实现安全控制的主流框架之一,提供了对资源访问的认证和授权功能。本任务将介绍 Spring Security 的基本概念、核心功能以及如何在 Spring Boot 项目中引入 Spring Security。

10.1.1　Security 简介

Spring Security 是 Spring 生态中一款开源的安全访问控制框架。它提供了一组可以在 Spring 应用上下文中配置的 Bean,利用 Spring IOC、DI 和 AOP 功能为 Spring 应用系统提供声明式的安全访问控制功能,减少了为系统安全控制编写大量重复代码的工作。

Spring Security 出现时间比 Spring Boot 还早。但是在早期的 Web 项目中应用 Spring Security 的配置过于烦琐,因此实际开发中 Spring Security 并不常用。自从有了 Spring Boot 框架之后,Spring Boot 框架对于 Spring Security 提供了自动化配置方案,使开发人员能够以最少的配置使用 Spring Security,Spring Security 的应用才日渐增多。

Spring Security 的核心功能就是认证和授权。认证就是对访问系统资源的身份信息进行核验,只有系统认定的合法用户才能访问系统资源,认证功能一般对应于系统资源的登录功能。授权就是当用户通过认证后,在访问系统资源时还要拥有对应的权限才能访问。如超级管理员能访问所有资源,普通用户只能访问部分资源。实现授权功能一般需要对各个用户设定不同的角色或资源访问权限。Spring Security 遵循先认证再授权的原则,先认证用户是否登录,如果已登录,则按照用户相应的角色或权限访问特定的资源。

在具体实现上，Spring Security 是通过一系列原生 Filter 组成过滤器链来拦截 Web 请求，Web 请求会一层层经过过滤器链的每个过滤器过滤，过滤完毕，完成资源访问的认证与授权。如果中间某个环节发现该请求未认证或者未授权，会抛出相应异常并交由异常处理器去处理。Spring Security 过滤器链如图 10-1 所示。

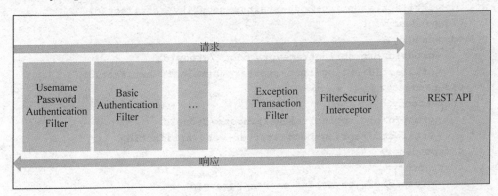

图 10-1　Spring Security 过滤器链

在图 10-1 中，Spring Security 过滤器链由一个个过滤器组成，每进入一个 Authentication Filter 时，过滤器就会根据请求判断是否认证。其中 FilterSecurity Interceptor 是过滤器的最后一层拦截，如果拦截通过，则放行进入 Restful API；如果拦截不通过，则将异常抛给异常处理器 Exception Transaction Filter 来处理。

10.1.2　Spring Boot 中引入 Spring Security

Spring Boot 能够很方便地引入 Spring Security，只需要在 pom.xml 文件里面引入如下 Spring Security 依赖即可。

```
<dependency>
  <groupId>org.springframework.boot</groupId>
  <artifactId>spring-boot-starter-security</artifactId>
</dependency>
```

引入完毕，Spring Boot 通过一系列 xxxAutoConfiguration 对 Spring Security 的一系列底层组件进行自动配置，开发人员可零配置使用 Spring Security。当然也可以根据实际需求修改默认配置。

任务 10.2　Spring Security 单用户认证和授权

本任务将介绍 Spring Security 的基本使用，包括默认登录注销认证编程、自定义登录注销认证编程、自定义授权编程和静态资源访问控制。

10.2.1　Spring Security 默认登录注销认证

这里新建一个项目 SpringDemo10，在项目的 pom.xml 文件中引入 web、thymeleaf 和

Spring Security 依赖。SpringDemo10 项目的 pom.xml 文件如下。

```xml
<dependencies>
    <dependency>
        <groupId>org.springframework.boot</groupId>
        <artifactId>spring-boot-starter-security</artifactId>
    </dependency>
    <dependency>
        <groupId>org.springframework.boot</groupId>
        <artifactId>spring-boot-starter-thymeleaf</artifactId>
    </dependency>
    <dependency>
        <groupId>org.springframework.boot</groupId>
        <artifactId>spring-boot-starter-web</artifactId>
    </dependency>
    <dependency>
        <groupId>org.thymeleaf.extras</groupId>
        <artifactId>thymeleaf-extras-springsecurity6</artifactId>
    </dependency>
    <dependency>
        <groupId>org.springframework.boot</groupId>
        <artifactId>spring-boot-starter-test</artifactId>
        <scope>test</scope>
    </dependency>
    <dependency>
        <groupId>org.springframework.security</groupId>
        <artifactId>spring-security-test</artifactId>
        <scope>test</scope>
    </dependency>
</dependencies>
```

项目前端使用 Thymeleaf 模板引擎，则需引入 thymeleaf-extras-springsecurityX 依赖以便 thymeleaf 支持 Spring Security 标签，目前最新版本为 6。

在 SpringDemo10 项目的 resources/templates 目录下新建项目首页 index.html。

【index.html】

```html
<!DOCTYPE html>
<html lang="en">
<head>
    <meta charset="UTF-8">
    <title>Title</title>
</head>
<body>
首页
</body>
</html>
```

在 SpringDemo10 项目的 springdemo10 目录下新建 controller 目录。在 controller 目录内部新建 MyController.java 文件，内部添加请求映射处理方法。

【MyController.java】

```
package springdemo10.controller;
import org.springframework.stereotype.Controller;
import org.springframework.web.bind.annotation.RequestMapping;
@Controller
public class MyController {
    @RequestMapping("/toIndex")
    public String toIndex(){
        return "index";
    }
}
```

启动 SpringDemo10 项目，可以看到控制台打印了一串随机字符串。

```
Using generated security password: 66cd9db3-98ec-44f8-b0c0-0e3aea169925
```

该随机字符串是 Spring Security 随机生成的密码。每次启动生成的密码都不一样。在浏览器地址栏中输入访问地址 http://localhost:8080/toIndex，页面并没有跳转到 index.html，而是跳转到 Spring Security 的默认登录页面，如图 10-2 所示。浏览器请求地址也变为 http://localhost:8080/login。/login 不是我们自己定义的，而是 Spring Security 默认跳转到登录页面的 URL。

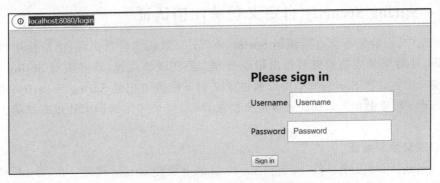

图 10-2　Spring Security 登录页面

实际上，Spring Security 将 SpringDemo10 内部的所有资源都保护起来了，无论访问 SpringDemo10 内部的哪个资源，页面都会先跳转到 Spring Security 的默认登录页面。只有在登录页面中输入正确的用户名和密码才能访问。Spring Security 默认提供的用户名为 user，密码为项目启动时打印的随机字符串密码。在图 10-3 中的登录页面中输入 Username 为 user，Password 为 66cd9db3-98ec-44f8-b0c0-0e3aea169925，单击 Sign in 按钮，页面跳转到 index.html，用户信息被保存在当前的 Session 中，后续操作都不需再次登录。

图 10-3　Spring Security 登录成功跳转 index.html

如果用户名和密码输入错误，页面不会跳转，并显示"用户名或密码错误"的提示信息。

在上述案例中，登录使用的用户名默认为 user，密码为随机字符串。如需指定特定的用户名和密码登录,可以在 SpringDemo10 项目的配置文件 application.yaml 中进行如下配置。如设置用户名为 admin,密码也为 admin。

```yaml
spring:
  security:
    user:
      name: admin
      password: admin
```

默认情况下,Spring Security 仅支持表单登录认证(FormLogin),且只能使用固定的用户名和随机生成的密码,不支持授权。如果要使用授权功能,需要进行一些额外配置。

SpringSecurity 中也提供了默认的注销功能,默认注销 URL 为/logout,注销时会清除认证信息和 Session 对象,注销完毕,系统默认跳转到登录页面。如使用默认注销功能,只需在页面中添加注销按钮,保证注销 URL 为/logout,请求方式为 post 即可。Spring Security 自带了 CSRF 防御机制,这里表单提交必须使用 post 请求,如使用 get 请求将报错。如在 index.html 页面中就可以添加如下代码实现注销功能。

```html
<form th:action="@{/logout}" method="post">
    <input type="submit" value="注销" />
</form>
```

10.2.2　Spring Security 自定义登录注销认证

在开发中,一般都不会直接使用 Spring Security 默认的登录页面,而是使用自定义的登录页面,并对登录成功或失败作出相应处理。要实现该功能,就必须对 Spring Security 进行自定义配置。在 Spring 6 中可以通过定义如下配置类配置 Spring Security,类中利用 HttpSecurity 对象的相关方法配置自定义登录界面、登录和注销请求处理方法等。具体实现如下。

1. 自定义登录认证

在 SpringDemo10 项目的 springdemo10 目录下新建 config 目录。在 config 目录内部新建 SecurityConfig.java 文件,用于配置 Spring Security。SecurityConfig 类上必须添加@Configuration 和@EnableWebSecurity 注解,以表示该类是一个配置类且已开启 Security 配置。SecurityConfig.java 文件内代码如下。

【SecurityConfig.java】

```java
package springdemo10.config;
import org.springframework.context.annotation.Bean;
import org.springframework.context.annotation.Configuration;
import org.springframework.security.config.annotation.web.builders.HttpSecurity;
import org.springframework.security.config.annotation.web.
        configuration.EnableWebSecurity;
import org.springframework.security.web.SecurityFilterChain;
@Configuration
@EnableWebSecurity
```

```java
public class SecurityConfig{
    @Bean
    public SecurityFilterChain filterChain(HttpSecurity http) throws Exception {
        //使用自定义登录页面认证并设定认证规则
        http.formLogin()
                .loginPage("/")                              //设置登录页面访问路径
                .loginProcessingUrl("/userLogin")            //设置登录处理请求URL
                .defaultSuccessUrl("/toIndex")               //设置登录成功处理请求URL
                .failureUrl("/?error=true")                  //设置登录失败处理请求URL
                .permitAll();                                //设置登录页面不需要认证访问
        //除此之外,任意请求必须认证后才可以访问
        http.authorizeHttpRequests(authorize ->authorize
                .anyRequest().authenticated());
        return http.build();
    }
}
```

修改 MyController.java 文件,内部添加 toLogin 方法。如果访问项目根目录或 toLogin,页面跳转到登录页面 login.html。

```java
@RequestMapping({"/","/toLogin"})
public String toLogin(){
    return "login";
}
```

SpringDemo10 项目的 resources/templates 目录下新建项目登录页 login.html。由于 SpringSecurity 的 CSRF 防御机制,这里表单提交使用 post 请求。

【login.html】

```html
<!DOCTYPE html>
<html lang="en" xmlns:th="http://www.thymeleaf.org">
<head>
    <meta charset="UTF-8">
    <title>Title</title>
</head>
<body>
<form th:action="@{/userLogin}" method="post">
<!--如果请求路径中包含 error 参数,则登录失败-->
    <div th:if="${param.error}">
        <span th:text="登录失败"></span>
    </div><br/>
    用户名:<input type="text" name="username" /><br/>
    密   码:<input type="text" name="password" /><br/>
    <input type="submit" value="提交" />
</form></body>
</html>
```

启动 SpringDemo10 项目,在浏览器地址栏输入访问地址 http://localhost:8080/,页面跳转到自定义登录页面,如图 10-4 所示。

在图10-4页面中输入用户名和密码,如果用户名和密码都为 admin,则登录成功跳转图10-3所示首页;如果用户名和密码输入错误,则页面跳转回 login.html 并显示登录失败信息,如图10-5所示。

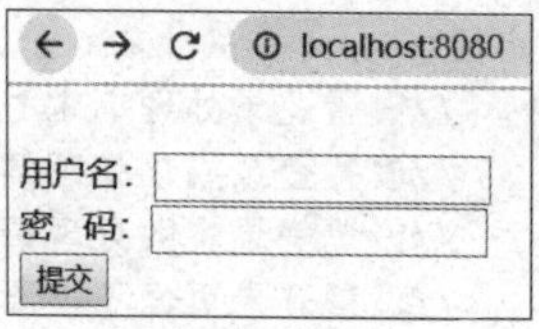

图10-4 自定义登录页面 login.html

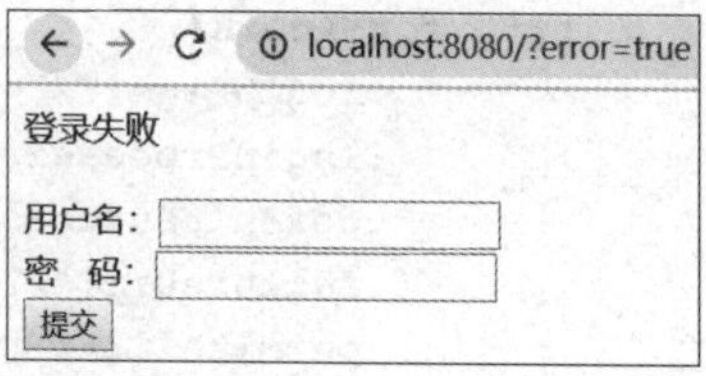

图10-5 登录失败

如果项目是前后端分离架构,页面跳转完全由前端控制,那么后端认证处理程序只需作为一个 API 接口返回登录成功或失败的信息给前端框架即可,前端框架根据结果自行进行页面跳转。这里就需要用到 successHandler 和 failureHandler 两个方法。successHandler 对应于登录成功处理,failureHandler 对应于登录失败处理。通过这两个方法就可以返回登录成功或失败的信息。具体修改 SecurityConfig 类内部的 filterChain 方法如下。

```java
@Bean
public SecurityFilterChain filterChain2(HttpSecurity http) throws Exception {
    //使用自定义登录页面认证并设定认证规则
    http.authorizeHttpRequests(authorize ->authorize
            .anyRequest().authenticated())         //任意请求必须认证后才可以访问
        .formLogin()
            .loginPage("/")                         //设置登录页面访问路径
            .loginProcessingUrl("/userLogin")       //设置登录处理请求 URL
            //设置登录成功处理请求类
            .successHandler((req, resp, authentication) -> {
                //principal 对象封装用户登录信息
                Object principal = authentication.getPrincipal();
                resp.setContentType("application/json;charset=utf-8");
                resp.getWriter().write(
                    new ObjectMapper().writeValueAsString(principal))
            })
            //设置登录失败处理请求类
            .failureHandler((req, resp, exp) -> {
                resp.setContentType("application/json;charset=utf-8");
                resp.getWriter().write(exp.getMessage());
            })
            .permitAll();                           //设置登录页面不需要认证访问
    return http.build();
}
```

上述代码使用 successHandler 方法和 failureHandler 方法执行登录成功和失败操作。successHandler 方法内部有三个参数,通常命名为 req、resp 和 authentication。其中,req 是 HttpServletRequest 对象,resp 是 HttpServletResponse 对象,authentication 封装了登录成功后的用户信息,可以通过 authentication 对象的 getPrincipal 方法获取登录成功后的用户

信息。failureHandler 方法内部也有三个参数,通常命名为 req、resp 和 exp。其中,req 是 HttpServletRequest 对象,resp 是 HttpServletResponse 对象,exp 封装了登录失败的原因。可以通过 exp 对象的 getMessage 方法获取登录失败的原因。

启动 SpringDemo10 项目,在浏览器地址栏中输入访问地址 http://localhost:8080/,页面跳转到自定义登录页面,如果登录成功,当前页面显示登录用户信息如下。

```
{"password":null,"username":"admin","authorities":[{"authority":"ROLE_admin"}],
"accountNonExpired":true,"accountNonLocked":true,"credentialsNonExpired":true,
"enabled":true}
```

这里 password 为 null,是因为用户输入的密码字符串在认证过后通常会被移除,以保障用户安全。authority 为 ROLE_admin,是因为在权限控制时,如果登录时传入的角色名称前缀不是 ROLE_,Spring Security,会自动把传入的角色名称拼接上 ROLE_前缀作为用户角色去进行匹配。如果直接使用权限进行匹配,则不改变原有权限数据。

如果登录失败,当前页面显示 Spring Security 默认的登录错误信息,用户名或密码错误。

2. 自定义注销认证

SpringSecurity 中提供了默认的注销功能。如不想使用默认注销功能,需要对注销功能进行自定义配置,可以在 SecurityConfig 配置类的方法中添加如下配置代码。

```
http.logout()
//注销登录请求地址,默认路径为/logout
.logoutUrl("/logout")
//是否使 session 失效,默认为 true
.invalidateHttpSession(true)
//是否清除认证信息,默认为 true
.clearAuthentication(true)
//注销登录后的跳转地址,一般为登录页面
.logoutSuccessUrl("/login.html");
```

如果项目是前后端分离架构,注销时页面跳转完全由前端控制,那么后端认证处理程序只需作为一个 API 接口返回注销成功或失败的信息给前端框架即可,前端框架根据结果自行进行页面跳转。可以利用 logoutSuccessHandler 方法实现注销处理。如在 SecurityConfig 配置类内部的 filterChain 方法添加如下代码,就可以在注销成功后向前端发送注销成功信息。

```
http.logout().logoutSuccessHandler((req, resp, authentication) -> {
    resp.setContentType("application/json;charset=utf-8");
    resp.getWriter().write("注销成功");
});
```

10.2.3 Spring Security 自定义授权

在用户认证成功登录系统后,一般会根据用户角色或权限设定用户能够访问的资源。

这就需要用到 Spring Security 的授权功能了。Spring Security 主要通过 Authority-AuthorizationManager 对象的 hasRole、hasAnyRole、hasAuthority 和 hasAnyAuthority 这 4 种方法进行权限控制。其中在 hasRole 和 hasAnyRole 方法中如果传入的是角色，且前缀不是 Role_，Spring Security 会为角色添加 Role_ 前缀；如果传入的是权限，则保持原样。Spring Security 权限控制方法如表 10-1 所示。

表 10-1 Spring Security 权限控制方法

方　　法	作　　用
hasRole	hasRole 接收一个字符串输入参数，要求用户只有一个角色，角色匹配才通过
hasAnyRole	hasAnyRole 接收多个字符串输入参数，支持用户具备多个角色，只要有一个角色匹配就通过。如用户同时具有普通用户角色和管理员角色
hasAuthority	hasAuthority 接收一个字符串输入参数，要求用户只有一个权限，权限匹配才通过
hasAnyAuthority	hasAnyAuthority 接收多个字符串输入参数，支持用户具备多个权限，只要有一个权限匹配就通过。如用户同时具有浏览 A 页面和 B 页面的权限

基于角色的权限控制在实际使用中更为常见，下面以 hasRole 方法为例介绍 Spring Security 的授权功能实现。在配置文件 application.yaml 中配置了用户信息如下：用户名和密码为 admin，角色为 admin。

```yaml
spring:
  security:
    user:
      name: admin
      password: admin
      roles:
        - admin
```

在 index.html 页面添加两个超链接，分别跳转到 admin.html 和 test.html 两个页面。用户登录 index.html 页面后，只有具备 admin 角色才可以访问 admin.html 页面，只有具备 test 角色才可以访问 test.html 页面。index.html、admin.html 和 test.html 页面代码如下。

【index.html】

```html
<!DOCTYPE html>
<html lang="en">
<head>
    <meta charset="UTF-8">
    <title>Title</title>
</head>
<body>
首页<br>
<a href="/toAdmin">toAdmin</a>
<a href="/toTest">toTest</a>
</body>
</html>
```

【admin.html】

```html
<!DOCTYPE html>
<html lang="en">
<head>
    <meta charset="UTF-8">
    <title>Title</title>
</head>
<body>
admin角色进入
</body>
</html>
```

【test.html】

```html
<!DOCTYPE html>
<html lang="en">
<head>
    <meta charset="UTF-8">
    <title>Title</title>
</head>
<body>
test角色进入
</body>
</html>
```

在编程使用中,Spring Security 的授权功能有两种实现方式:一种是在配置类内部进行配置,另一种是利用注解实现。这里假定项目前后端不分离,分别介绍二者的具体实现。

1. 在配置类内部进行配置

直接修改 SecurityConfig 配置类内部代码如下,内部添加 hasRole 方法进行权限控制。

```java
@Bean
public SecurityFilterChain filterChain1(HttpSecurity http) throws Exception {
    //使用自定义登录页面认证并设定认证规则
    http.authorizeHttpRequests(authorize -> authorize
            //admin.html需认证后且用户拥有admin角色才能访问
            .requestMatchers("/toAdmin").hasRole("admin")
            //test.html需认证后且用户拥有test角色才能访问
            .requestMatchers("/toTest").hasRole("test")
            .anyRequest().authenticated())          //任意请求必须认证后才可以访问
            .formLogin()
            .loginPage("/")                          //设置登录页面访问路径
            .loginProcessingUrl("/userLogin")        //设置登录处理请求URL
            .defaultSuccessUrl("/toIndex")           //设置登录成功处理请求URL
            .failureUrl("/?error=true")              //设置登录失败处理请求URL
            .permitAll();                            //设置登录页面不需要认证访问
    return http.build();
}
```

启动 SpringDemo10 项目，在浏览器地址栏中输入访问地址 http://localhost:8080/，页面跳转到自定义登录页面，如果登录成功，跳转到 index.html 页面，如图 10-6 所示。

图 10-6　index.html 页面

在图 10-6 所示页面中，如果单击 toAdmin 超链接，由于登录用户角色为 admin，能够正常进入 admin.html 页面，如图 10-7 所示。如果单击 toTest 超链接，由于登录用户角色不是 test，页面跳转到默认的 403 页面，如图 10-8 所示。

图 10-7　进入 admin 页面

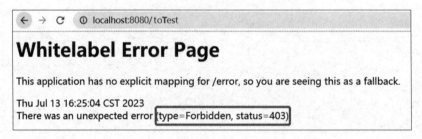

图 10-8　进入默认的 403 页面

2. 利用注解实现

在项目开发中，利用注解实现权限控制更为常用。使用注解进行权限控制首先需开启相关配置。在 Spring Security 配置类上添加如下注解。

```
@Configuration
@EnableMethodSecurity
public class SecurityConfig{
}
```

@EnableMethodSecurity 注解使我们可以直接在控制器类或控制器类内部的映射请求处理方法上添加 @PreAuthorize 和 @PostAuthorize 两个注解进行资源访问认证。其中 @PreAuthorize 注解会在执行前进行认证，而 @PostAuthorize 注解会在执行后进行认证。同时在 @PreAuthorize 和 @PostAuthorize 注解内部可利用 hasRole、hasAnyRole、hasAuthority 和 hasAnyAuthority 等方法进行权限控制。

例如，将配置类 SecurityConfig 内部如下两条权限控制代码删除。

```
.requestMatchers("/toAdmin").hasRole("admin")
.requestMatchers("/toTest").hasRole("test")
```

在 MyController 类的 toAdmin 方法和 toTest 方法上添加如下粗体代码，同样能实现

图 10-6 和图 10-7 所示的权限控制功能。

```
@PreAuthorize("hasRole('admin')")
@RequestMapping("/toAdmin")
public String toAdmin(){
    return "admin";
}
@PreAuthorize("hasRole('test')")
@RequestMapping("/toTest")
public String toTest(){
    return "test";
}
```

注意：配置文件 application.yaml 中配置的角色为 admin，但是在权限匹配时，Spring Security 会自动为 application.yaml 中配置的角色添加 ROLE_ 前缀，同时 hasRole 方法中定义的角色也会被添加 ROLE_ 前缀。

如果项目采用前后端分离架构，用户没有访问权限，后端将作为一个 API 接口将向前端返回没有权限等信息。前端接收到没有权限的信息后，再控制页面跳转默认的 403 页面。具体实现上可在 Spring Security 配置类中利用 accessDeniedHandler 方法实现，accessDeniedHandler 方法内部有三个参数，通常命名为 req、resp 和 exp。其中，req 是 HttpServletRequest 对象；resp 是 HttpServletResponse 对象；exp 封装了用户访问资源失败的原因，可以使用 exp 的 getMessage 方法获取访问资源失败原因。具体使用如下：修改 SecurityConfig 类内部的 filterChain 方法如下。

```
@Bean
public SecurityFilterChain filterChain3(HttpSecurity http) throws Exception {
    //使用自定义登录页面认证并设定认证规则
    http.authorizeHttpRequests(authorize -> authorize
        //admin.html 需认证后且用户拥有 admin 角色才能访问
        .requestMatchers("/toAdmin").hasRole("admin")
        //test.html 需认证后且用户拥有 test 角色才能访问
        .requestMatchers("/toTest").hasRole("test")
        .anyRequest().authenticated())            //任意请求必须认证后才可以访问
        .formLogin()
        .loginPage("/")                            //设置登录页面访问路径
        .loginProcessingUrl("/userLogin")          //设置登录处理请求 URL
        .defaultSuccessUrl("/toIndex")             //设置登录成功处理请求 URL
        .failureUrl("/?error=true")                //设置登录失败处理请求 URL
        .permitAll();                              //设置登录页面不需要认证访问
    //处理没有权限访问资源的情况
    http.exceptionHandling().accessDeniedHandler((req, resp, exp) -> {
        resp.setContentType("application/json;charset=utf-8");
        resp.getWriter().write(exp.getMessage());
    });
    return http.build();
}
```

启动 SpringDemo10 项目，使用 admin 用户登录成功后访问 test.html 页面。由于没有权限访问，页面显示访问失败信息为 Access Denied。

10.2.4 Spring Security 静态资源的访问控制

Web 应用开发中少不了会引入一些 JS、CSS 和图片等静态资源美化页面。如果页面中引入了静态资源，但是又不想 Spring Security 对静态资源进行认证拦截，可以在配置类 SecurityConfig 中进行如下配置。这里假定在项目的 resources/static 目录下存在 css 目录，内部存放多个 CSS 文件。

```
@Bean
public SecurityFilterChain filterChain3(HttpSecurity http) throws Exception {
    http.authorizeHttpRequests(authorize -> authorize
            .requestMatchers("/css/**").permitAll()    //放行 css 目录下的所有文件
            .anyRequest().authenticated())
}
```

如果需要放行 JS、图片等静态资源，也可以参考上述代码。

任务 10.3　Spring Security 多用户认证和授权

在任务 10.2 中只创建了一个用户进认证和授权。但在实际使用中，一般会有多个用户存在，不同用户的用户名、密码和权限都不相同。如何创建多个用户呢？一种方式就是在内存中创建，另一种方式就是在数据库中创建，在实际开发中后一种方式更为常见。下面将介绍基于数据库实现多用户认证和授权。

基于数据库实现多用户认证和授权时，创建的用户信息不再写死在程序代码中，而是从数据库中实时读取出来的。要实现此功能，需自定义一个 Java 类并继承 UserDetailsService 接口，类内部覆写 loadUserByUsername 方法，从数据库中读取用户信息，最终创建 UserDetails 用户对象，然后将该 Java 类使用 auth.userDetailsService 方法添加进 AuthenticationManagerBuilder 对象中。

创建 springdemo10 数据库中存在 userinfo 数据表，userinfo 数据表中的数据如表 10-2 所示。密码先不采用加密处理，在数据库中保存的密码前面加{noop}。

表 10-2　userinfo 数据表中的数据

id	username	password	roles
1	admin	{noop}admin	admin
2	test	{noop}test	test

通过 Mybatis 读取数据库用户信息，用户信息为 UserDomain 对象，内部包含用户名、密码和角色三个属性。Mybatis 映射接口 UserMapper 已提供用户信息查询方法 selectuserbyusername。具体实现代码如下。

【MyUserDetailsService】

```java
package springdemo10.config;
import org.springframework.beans.factory.annotation.Autowired;
import org.springframework.security.core.authority.SimpleGrantedAuthority;
import org.springframework.security.core.userdetails.User;
import org.springframework.security.core.userdetails.UserDetails;
import org.springframework.security.core.userdetails.UserDetailsService;
import org.springframework.security.core.userdetails.UsernameNotFoundException;
import org.springframework.stereotype.Service;
import springdemo10.domain.UserDomain;
import springdemo10.mapper.UserMapper;
import java.util.ArrayList;
import java.util.List;
@Service
public class MyUserDetailsService implements UserDetailsService {
    @Autowired
    UserMapper userMapper;
    @Override
    public UserDetails loadUserByUsername(String username)
            throws UsernameNotFoundException {
        //此处可以调用数据库查询方法且通过用户名查询该用户信息
        UserDomain user = userMapper.selectuserbyusername(username);
        if (user == null) {
            throw new UsernameNotFoundException("user not found");
        }
        //权限可以有多个,可以是一个 SimpleGrantedAuthority 对象集合
        List<SimpleGrantedAuthority> authorities =
                new ArrayList<SimpleGrantedAuthority>();
        //如传入角色,角色必须手动添加前缀 ROLE_;如果有多个角色,可循环加入
        authorities.add(
                new SimpleGrantedAuthority("ROLE_"+user.getRoles()));
        //利用查询的用户信息和 authorities 构建 UserDetails 对象并返回外界
        UserDetails userDetails=
                new User(user.getUsername(),user.getPassword(),authorities);
        return userDetails;
    }
}
```

注意:上述代码中角色为外界数据库中获取,需自行添加 ROLE_前缀。

在 SecurityConfig 配置类内部将 MyUserDetailsService 对象添加进 AuthenticationManagerBuilder 对象中。

```java
package springdemo10.config;
import org.springframework.beans.factory.annotation.Autowired;
import org.springframework.context.annotation.Bean;
import org.springframework.context.annotation.Configuration;
import org.springframework.security.config.annotation.authentication
        .builders.AuthenticationManagerBuilder;
```

```java
import org.springframework.security.config.annotation
        .method.configuration.EnableMethodSecurity;
import org.springframework.security.config.annotation.web.builders.HttpSecurity;
import org.springframework.security.config.annotation
        .web.configuration.EnableWebSecurity;
import org.springframework.security.web.SecurityFilterChain;
@Configuration
@EnableWebSecurity
@EnableMethodSecurity
public class SecurityConfigFromDB {
    //注入 MyUserDetailsService 对象
    @Autowired
    private MyUserDetailsService myUserDetailsService;
    @Bean
    public SecurityFilterChain filterChain(HttpSecurity http) throws Exception {
        //使用自定义登录页面认证并设定认证规则
        http.authorizeHttpRequests(authorize -> authorize
            //admin.html 需认证后且用户拥有 admin 角色才能访问
            .requestMatchers("/toAdmin").hasRole("admin")
            //test.html 需认证后且用户拥有 test 角色才能访问
            .requestMatchers("/toTest").hasRole("test")
            .anyRequest().authenticated())       //任意请求必须认证后才可以访问
            .formLogin()
            .loginPage("/")//设置登录页面的访问路径
            .loginProcessingUrl("/userLogin")    //设置登录处理请求 URL
            .defaultSuccessUrl("/toIndex")       //设置登录成功处理请求 URL
            .failureUrl("/?error=true")          //设置登录失败处理请求 URL
            .permitAll();                        //设置登录页面不需要认证访问
        http.logout().logoutUrl("/logout").logoutSuccessUrl("/");
        return http.build();
    }
    //向 AuthenticationManagerBuilder 对象添加 MyUserDetailsService 对象
    @Autowired
    public void configure(AuthenticationManagerBuilder auth) throws Exception{
        auth.userDetailsService(myUserDetailsService);
    }
}
```

如果数据库中保存的密码前面没有加{noop}，Spring Security 默认需要对密码进行加密操作。如采用 BCryptPasswordEncoder 加密，可以在 MyUserDetailsService 内对密码 password 作如下处理。

```java
String password=new BCryptPasswordEncoder().encode(user.getPassword());
        UserDetails userDetails=
                new User(user.getUsername(),password,authorities);
return userDetails;
```

然后在 SecurityConfig 的 configure 方法中为 AuthenticationManagerBuilder 对象再添加 BCryptPasswordEncoder 对象用于加密认证。

```
@Autowired
    public void configure(AuthenticationManagerBuilder auth)throws Exception{
        auth.userDetailsService(myUserDetailsService)
            .passwordEncoder(new BCryptPasswordEncoder());
    }
```

任务 10.4　综合案例：利用 Spring Security 进行安全控制

下面以一个综合案例演示如何利用 Spring Security 对项目 9 的 SpringDemo9_Thymeleaf 项目进行用户认证和授权。

10.4.1　案例任务

将 springdemo9 数据库中的 employee 表复制到 springdemo10 数据库。修改 springdemo10 数据库内部用户信息表 userinfo 的数据，去除 admin 用户和 test 用户密码前面的{noop}，最终 userinfo 内部数据如表 10-3 所示。

表 10-3　修改后 userinfo 数据表的数据

id	username	password	roles
1	admin	admin	admin
2	test	test	test

为 SpringDemo9_Thymeleaf 项目添加登录页面，登录成功跳转到查询员工页面 select_emp.html，登录失败跳转回登录页面并显示错误信息。如果登录成功，则通过登录用户角色进行访问控制，admin 角色可以访问所有资源，test 角色只能查询员工信息，后台与数据库交互需使用 Mybatis 框架。

10.4.2　案例分析

SpringDemo9_Thymeleaf 是一个使用 Thymeleaf 模板引擎的 Spring Boot 项目，案例编程时需考虑如下因素。

（1）Spring Security 配置类中需自定义登录成功和登录失败的页面跳转逻辑。

（2）用户信息从数据库中读取需自定义 Java 类并继承 UserDetailsService 接口，内部覆写 loadUserByUsername 方法来读取数据库数据。由于用户 1 和用户 2 的密码前都没有加{noop}，在 loadUserByUsername 方法中需要对密码进行加密处理，可采用 BCryptPasswordEncoder 加密。

（3）为方便，可使用@PreAuthorize 注解实现权限控制。

10.4.3　案例实施

由于篇幅有限，任务实施具体步骤采用二维码形式展示。

综合案例的实施步骤

小 结

本项目中详细介绍了 Spring Security 的认证授权功能，包括 Spring Security 基本概念、Spring Security 默认认证授权编程实现、Spring Security 自定义单用户认证授权编程实现、Spring Security 基于内存和数据库的多用户认证授权编程实现，使读者进一步熟悉了 Spring Security 的安全控制机制和编程应用。最后通过一个综合案例演示如何在 Spring Boot+Thymeleaf 项目中结合 MySQL 数据库实现不同用户、不同角色对资源访问的认证和授权。

课后练习：前后端分离项目的安全控制

利用 Spring Security 对项目 9 中任务 9.5 的综合案例的前后端分离项目——员工信息管理进行安全控制，要求 admin 用户可以访问所有功能，test 用户只能查询员工信息。

项目 11　Spring Boot Web 项目部署

项目开发、测试完毕，还需要进行上线部署。在之前的项目中已经建立了很多 Spring Boot 项目，本项目将基于项目 10 和项目 9 介绍 Spring Boot Web 项目前后端不分离架构和前后端分离架构下的项目部署。

任务 11.1　部署前后端不分离项目

在 Web 项目前后端不分离架构下，只有一个 Spring Boot 项目，项目中前后端内容整合在一起，部署起来也相对简单。在项目部署前通常需要将项目打包，主要打包成 Jar 项目和 War 项目两种方式，这两种方式部署的步骤是不一样的。这里将分别介绍 Windows 环境下 Spring Boot 基于 Jar 项目和 War 项目的两种部署方式。

11.1.1　基于 Jar 项目部署

下面以 SpringDemo10_Security 项目为例，介绍 Spring Boot Web 项目基于 Jar 的打包和部署。如果 Spring Boot 构建的是一个普通的 Java 服务端项目而不是 Web 项目，此部署方法仍然适用。

1. 检查是否引入 Spring Boot 打包插件

Spring Boot 的打包通过 <build> 标签下的打包插件 spring-boot-maven-plugin 来完成，默认打包成 Jar 项目。spring-boot-maven-plugin 内部可以定义一些打包的配置项。一般在项目正常创建完毕，pom.xml 文件中就会自动生成 <build> 标签，<build> 标签内部有类似如下的配置，如这里就设置了打包时排除对 lombok 插件进行打包。

```xml
<build>
    <plugins>
        <plugin>
            <groupId>org.springframework.boot</groupId>
            <artifactId>spring-boot-maven-plugin</artifactId>
            <configuration>
                <excludes>
                    <exclude>
                        <groupId>org.projectlombok</groupId>
                        <artifactId>lombok</artifactId>
                    </exclude>
                </excludes>
```

```
            </configuration>
        </plugin>
    </plugins>
</build>
```

2. 使用 maven package 指令对项目进行打包

在 idea 右侧的 Maven 操作栏中双击 package 选项，对项目进行打包。如果之前该项目已进行过打包，项目的 target 目录下已有内容，可先双击 clean 选项清除项目的 target 目录，然后双击 compile 选项对项目进行编译，最后双击 package 选项对项目进行重新打包，相关操作如图 11-1 所示。

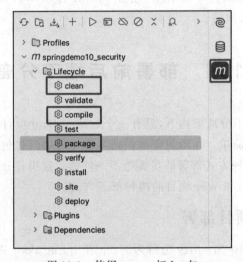

图 11-1 使用 maven 打 Jar 包

等待一会儿，如果控制台出现 Build SUCCESS 字样，则打包成功。打包完毕，生成的 Jar 包位于项目的 target 目录下，如图 11-2 所示。

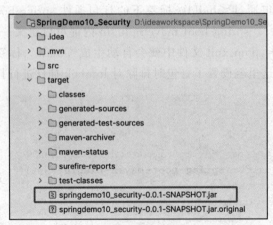

图 11-2 Jar 包位置

3. 运行 Jar 包

Spring Boot3.x 版本项目内置 Tomcat 10 服务器，不需借助外部 Web 服务器运行项目打包成功后,可输入 cmd 进入命令提示符界面,在命令提示符界面中进入 jar 包所在目录,输入如下命令运行 Jar 包。

```
java -jar springdemo10_security-0.0.1-SNAPSHOT.jar
```

待项目启动后,在浏览器地址栏中输入访问地址 http://localhost:8080/,页面如正常跳转到登录页面,则项目部署成功。

11.1.2 基于 War 项目部署

默认情况下,基于 Spring Boot 开发的 Web 项目作为一个 jar 项目部署运行,但是某些特殊情况下,如果要将 Web 项目放在外部的 Web 容器(例如 Tomcat)中部署运行,就需要将项目打包成 War 项目。下面将介绍 Spring Boot War 项目基于 Tomcat 服务器的部署。

本书使用的 Spring Boot 为 Spring Boot3.x 版本,开发环境默认支持的 Tomcat 版本为 Tomcat 10。为与开发环境保持一致,部署时使用的 Tomcat 为 Tomcat 10.1.11 版本。部署之前需先下载并安装 Tomcat 10.1.11,确保环境变量 JAVA_HOME 值为 Java 17 的安装目录。

Spring Boot 默认打包成 Jar 项目,如要打包成 War 项目,需对原有项目部分配置做修改,具体修改如下。

1. 修改 pom.xml 文件

在 pom.xml 文件中添加 War 包相关配置,并设置 spring-boot-starter-web 模块部署时排除内置 Tomcat。pom.xml 文件中修改内容如下。

```xml
<?xml version="1.0" encoding="UTF-8"?>
<project xmlns="http://maven.apache.org/POM/4.0.0"
    xmlns:xsi="http://www.w3.org/2001/XMLSchema-instance"
    xsi:schemaLocation="http://maven.apache.org/POM/4.0.0
    https://maven.apache.org/xsd/maven-4.0.0.xsd">
    <modelVersion>4.0.0</modelVersion>
    <!--打 War 包需添加配置-->
    <packaging>war</packaging>
    //此处省略代码
    <dependencies>
    <!--注释项目中自带的所有 Tomcat 插件在部署时不打包-->
        <!--<dependency>-->
            <!--<groupId>org.springframework.boot</groupId>-->
            <!--<artifactId>spring-boot-starter-web</artifactId>-->
        <!--</dependency>-->
        <dependency>
            <groupId>org.springframework.boot</groupId>
            <artifactId>spring-boot-starter-web</artifactId>
            <exclusions>
                <exclusion>
                    <groupId>org.springframework.boot</groupId>
                    <artifactId>spring-boot-starter-tomcat</artifactId>
                </exclusion>
            </exclusions>
        </dependency>
```

```xml
        <dependency>
            <groupId>org.springframework.boot</groupId>
            <artifactId>spring-boot-starter-tomcat</artifactId>
            <scope>provided</scope>
        </dependency>
        //此处省略代码
    </dependencies>
    <build>
        <!--设置打 War 包的包名,Tomcat 中项目访问地址为
        http://localhost:8080/springdemo10security-->
        <finalName>springdemo10security</finalName>
        //此处省略代码
    </build>
</project>
```

上述代码添加＜packaging＞标签并设置了打包类型为 War 项目,打包名称为 springdemo10security,同时在部署时去除了原有 Web 模块中的 Tomcat 插件。

2. 修改 Spring Boot 项目启动类

修改原有的 Spring Boot 的启动类,使之继承 SpringBootServletInitializer 类,并重写内部的 configure 方法。

```java
package springdemo10_security;
import org.springframework.boot.SpringApplication;
import org.springframework.boot.autoconfigure.SpringBootApplication;
import org.springframework.boot.builder.SpringApplicationBuilder;
import org.springframework.boot.web.servlet.support.SpringBootServletInitializer;
@SpringBootApplication
public class Springdemo10SecurityApplication extends SpringBootServletInitializer {
    public static void main(String[] args) {
        SpringApplication.run(Springdemo10SecurityApplication.class, args);
    }
    @Override
    protected SpringApplicationBuilder configure(SpringApplicationBuilder builder) {
        return builder.sources(Springdemo10SecurityApplication.class);
    }
}
```

3. 使用 maven package 指令对项目进行打包

这一步骤操作和打 Jar 包类似,在 idea 右侧的 Maven 操作栏中,双击"package"选项对项目进行打包。如之前该项目已进行过打包,项目的 target 目录下已有内容。可先双击"clean"选项清除项目的 target 目录,然后双击"compile"选项对项目进行编译,最后双击"package"选项对项目进行重新打包。

4. 复制 War 包到 Tomcat 的 webapps 目录下

等待一会儿,如果控制台出现 Build SUCCESS 字样,则打包成功。打包完毕,生成的 War 包位于项目的 target 目录下,如图 11-3 所示。

确保计算机已安装 Tomcat 10.1.11,将 War 包复制到 Tomcat 10.1.11 的 webapps 目录

下,如图 11-4 所示。

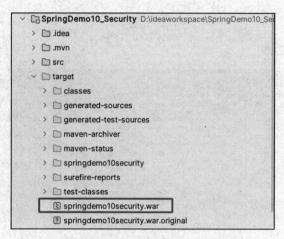

图 11-3　War 包位置

图 11-4　War 包复制到 Tomcat webapps 目录

5. 启动 Tomcat 并运行 War 包

在启动 Tomcat 之前还需配置一下日志输出编码格式,由于 Windows 下的 CMD 的默认编码是 GBK,Tomcat 控制台默认输出设置为 UTF-8 编码,启动后 Tomcat 命令行中文会出现乱码。可修改 Tomcat 安装目录下/conf/logging.properties 文件。将其中的

```
java.util.logging.ConsoleHandler.encoding=UTF-8
```

改成

```
java.util.logging.ConsoleHandler.encoding=GBK
```

在 Tomcat 10.1.11 的 bin 目录下单击 startup.bat 文件,启动 Tomcat,启动成功的界面如图 11-5 所示。

此时 Tomcat 会自动解压 webapps 目录下的 springdemo10security.war 文件。待 Tomcat 启动完毕,在浏览器地址栏中输入访问地址 http://localhost:8080/springdemo10security,页面如正常跳转到登录页面,则项目部署成功。如需停止 Tomcat,单击 Tomcat 的 bin 目录下的 shutdown.bat 文件。

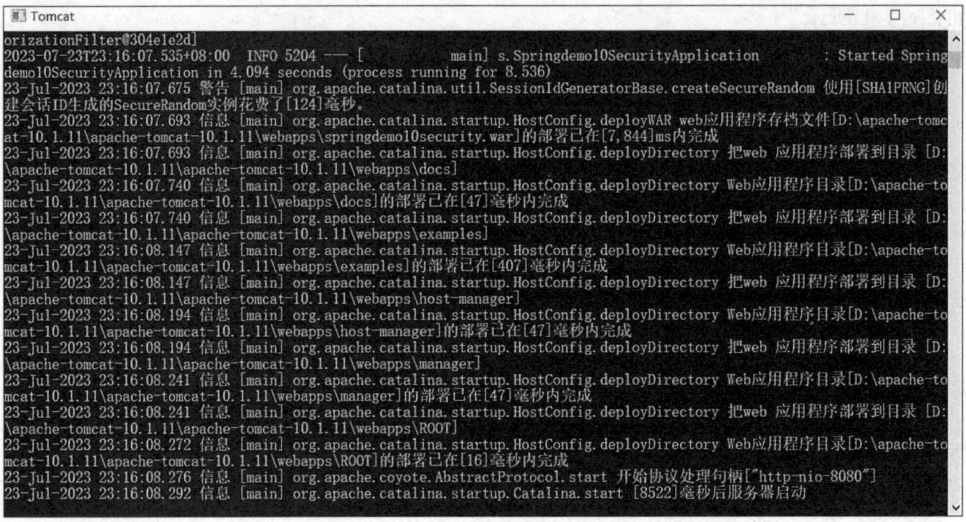

图 11-5 启动 Tomcat

任务 11.2 前后端分离项目部署

如果项目采用 Spring Boot＋Vue 的前后端分离架构，部署时就需要前后端项目分开部署，项目的后端 Spring Boot 作为服务端可参考 11.1.1 小节内容单独打包成 Jar 项目，利用 java -jar 命令运行提供后端服务，这里不再过多介绍。本任务主要介绍项目前端 Vue 项目的打包部署。

前后端分离项目不可避免会遇到跨域问题，在基于 Vite＋Vue3 的架构项目开发中可以通过在 vite.config.js 文件的 proxy 属性内部添加如下代码解决跨域问题。

```
server: {
    open: true,
    host: 'localhost',           //前端项目的 IP 地址
    port: 5173,                  //前端项目的端口号
    proxy: {
      //此处可以根据需求添加多个代理，分别映射不同的请求地址
      '/empApi': {
        target: 'http://localhost:8080',//后端项目的访问地址
        ws: true,
        changeOrigin: true,
        //替换 target 中的请求地址，
        //将 Axios 中的请求地址 http://localhost:8080/xxx 替换为 /empApi/xxx
        rewrite: (path) => path.replace(/^\/empApi/, '')
      }
    }
}
```

但上述 proxy 代理配置只在项目开发环境中有效，在项目打包部署运行时是无效的。为了使项目打包部署后也能实现跨域访问，就需要使用其他的代理工具，最常用的代理工具

就是 Nginx。Nginx 除了能解决跨域问题外，还能缓存静态资源，减少服务器压力，分发请求，实现服务器负载均衡。因此在实际应用中，Vue 项目大多采用 Nginx 部署。

由于本书的 Vue 项目采用 Vite+Vue3 架构，部署运行对浏览器版本有一定要求，要求 Chrome 浏览器版本大于或等于 87，Firefox 浏览器版本大于或等于 78，Safari 浏览器版本大于或等于 14，Edge 浏览器版本大于或等于 88。如浏览器版本过低，Vue 项目部署后访问会出现页面空白。

假设后端服务 SpringDemo9_Thymeleaf 已通过 maven 工具打包完成，如图 11-6 所示。

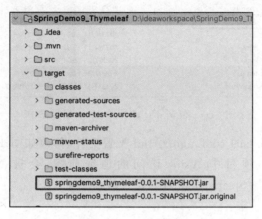

图 11-6 maven 打包

在命令提示符界面中进入 jar 包所在目录，输入如下命令，运行 Jar 包启动后端服务。

```
java -jar springdemo9_thymeleaf-0.0.1-SNAPSHOT.jar
```

下面详细介绍前端 Vue 项目的部署步骤。

1. 打包 Vue 项目

打开 VsCode，进入 vue001 项目根目录，并利用 npm run build 命令进行打包，如图 11-7 所示。

等待一会儿，待打包完成后，在 Vue001 项目的根目录下会生成 dist 目录，如图 11-8 所示。dist 目录内部就是 Vue 项目打包后的内容，其中 index.html 是 Vue 项目的入口文件，所有被使用到的 Vue 文件被打包成一个个 JS 文件。

图 11-7 Vue 项目打包

图 11-8 dist 目录内容

2. 安装并配置 Nginx

先从 Nginx 官网下载 Windows 版本的 Nginx 压缩包文件，将该压缩包文件解压即可使用。这里采用 Nginx 1.24.0 版本。将 Vue001 项目的 dist 目录复制到 Nginx 根目录下，如图 11-9 所示。

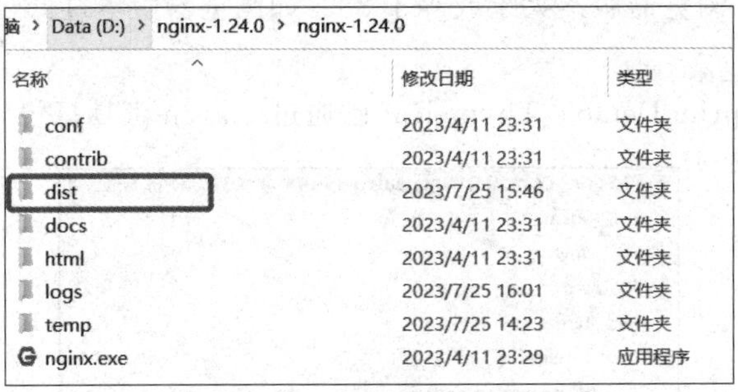

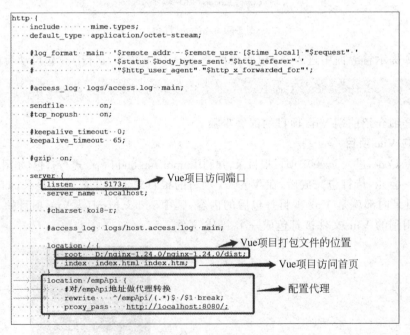

图 11-9　复制 dist 目录到 nginx 根目录中

修改 Nginx 根目录下的 conf/nginx.conf 配置文件，修改如图 11-10 中方框所示内容。其中，/empApi 是 Vue 项目中 Axios 访问的地址前缀，会被 Nginx 代理成 http://localhost:8080/并与后端 Spring Boot 项目进行数据交互。

图 11-10　修改 nginx.conf 配置文件

3. 启动 Nginx

在 Nginx 根目录下打开 CMD 命令提示符窗口，在其中输入 start nginx 命令启动 Nginx，然后在浏览器地址栏中输入访问地址 http://localhost:5173/，如果页面能够正常显示所有员工信息列表，则项目部署成功。如需关闭 Nginx，在命令提示符窗口输入 nginx.exe -s stop 命令即可。

小 结

本项目详细介绍了 Spring Boot Web 项目的部署,包括前后端不分离架构下 Spring Boot Thymeleaf 项目基于 Jar 项目和 War 项目的打包部署,前后端分离架构下 Spring Boot ＋ Vue＋Nginx 项目的打包部署,使读者进一步掌握了 Spring Boot Web 项目多种不同的打包部署方式。

课后练习:学生信息管理项目部署

对项目 9 任务 9.7 的课后练习 Spring Boot＋Vue 学生信息管理项目进行部署,部署时需使用 Nginx 工具。

参 考 文 献

[1] 丁雪丰. 学透 Spring：从入门到项目实战[M]. 北京：人民邮电出版社，2023.
[2] 郝佳. Spring 源码深度解析[M]. 2 版. 北京：人民邮电出版社，2019.
[3] 周红亮. Spring Boot 3 核心技术与最佳实践[M]. 北京：电子工业出版社，2023.
[4] 朱建昕. Spring Boot 3＋Vue 3 开发实战[M]. 北京：电子工业出版社，2023.
[5] 李西明，陈立为. Spring Boot 3.0 开发实战[M]. 北京：清华大学出版社，2023.
[6] 陈恒，等. SSM ＋ Spring Boot ＋ Vue.js 3 全栈开发从入门到实战(微课视频版)[M]. 北京：清华大学出版社，2022.
[7] 陈学明. Spring＋Spring MVC＋MyBatis 整合开发实战[M]. 北京：机械工业出版社，2020.
[8] 王松. Spring Boot＋Vue 全栈开发实战[M]. 北京：清华大学出版社，2018.
[9] 唐文. Spring Boot＋Vue 前后端分离项目全栈开发实战[M]. 北京：中国铁道出版社，2023.
[10] 肖海鹏，等. Spring Framework 6 开发实战(Spring＋Spring Web MVC＋MyBatis)[M]. 北京：清华大学出版社，2023.
[11] 朱智胜. Spring Boot 技术内幕：架构设计与实现原理[M]. 北京：机械工业出版社，2020.
[12] 龙中华. Spring Boot 实战派[M]. 北京：电子工业出版社，2020.
[13] 颜井赞. Spring Boot 整合开发案例实战[M]. 北京：清华大学出版社，2023.
[14] 张科. Spring Boot 企业级项目开发实战[M]. 北京：机械工业出版社，2022.